SpringerBriefs in Computer Science

Series Editors
Stan Zdonik
Peng Ning
Shashi Shekhar
Jonathan Katz
Xindong Wu
Lakhmi C. Jain
David Padua
Xuemin Shen
Borko Furht
V.S. Subrahmanian
Martial Hebert
Katsushi Ikeuchi
Bruno Siciliano

T0214306

For further volumes:
http://www.springer.com/series/10028

Shamanth Kumar • Fred Morstatter • Huan Liu

Twitter Data Analytics

 Springer

Shamanth Kumar
Data Mining and Machine Learning Lab
Arizona State University
Tempe, AZ, USA

Fred Morstatter
Data Mining and Machine Learning Lab
Arizona State University
Tempe, AZ, USA

Huan Liu
Data Mining and Machine Learning Lab
Arizona State University
Tempe, AZ, USA

ISSN 2191-5768   ISSN 2191-5776 (electronic)
ISBN 978-1-4614-9371-6   ISBN 978-1-4614-9372-3 (eBook)
DOI 10.1007/978-1-4614-9372-3
Springer New York Heidelberg Dordrecht London

Library of Congress Control Number: 2013953291

© The Author(s) 2014
This work is subject to copyright. All rights are reserved by the Publisher, whether the whole or part of the material is concerned, specifically the rights of translation, reprinting, reuse of illustrations, recitation, broadcasting, reproduction on microfilms or in any other physical way, and transmission or information storage and retrieval, electronic adaptation, computer software, or by similar or dissimilar methodology now known or hereafter developed. Exempted from this legal reservation are brief excerpts in connection with reviews or scholarly analysis or material supplied specifically for the purpose of being entered and executed on a computer system, for exclusive use by the purchaser of the work. Duplication of this publication or parts thereof is permitted only under the provisions of the Copyright Law of the Publisher's location, in its current version, and permission for use must always be obtained from Springer. Permissions for use may be obtained through RightsLink at the Copyright Clearance Center. Violations are liable to prosecution under the respective Copyright Law.
The use of general descriptive names, registered names, trademarks, service marks, etc. in this publication does not imply, even in the absence of a specific statement, that such names are exempt from the relevant protective laws and regulations and therefore free for general use.
While the advice and information in this book are believed to be true and accurate at the date of publication, neither the authors nor the editors nor the publisher can accept any legal responsibility for any errors or omissions that may be made. The publisher makes no warranty, express or implied, with respect to the material contained herein.

Printed on acid-free paper

Springer is part of Springer Science+Business Media (www.springer.com)

This effort is dedicated to my family. Thank you for all your support and encouragement. – SK

For my parents and Rio. Thank you for everything. – FM

To my parents, wife, and sons. – HL

Acknowledgements

We would like to thank the following individuals for their help in realizing this book. We would like to thank Daniel Howe and Grant Marshall for helping to organize the examples in the book, Daria Bazzi and Luis Brown for their help in proofreading and suggestions in organizing the book, and Terry Wen for preparing the web site. We appreciate Dr. Ross Maciejewski's helpful suggestions and guidance as our data visualization mentor. We express our immense gratitude to Dr. Rebecca Goolsby for her vision and insight for using social media as a tool for Humanitarian Assistance and Disaster Relief. Finally, we thank all members of the Data Mining and Machine Learning lab for their encouragement and advice throughout this process.

This book is the result of projects sponsored, in part, by the Office of Naval Research. With their support, we developed TweetTracker and TweetXplorer, flagship projects that helped us gain the knowledge and experience needed to produce this book.

Contents

Chapter 1
Introduction

Twitter[®][1] is a massive social networking site tuned towards fast communication. More than 140 million active users publish over 400 million 140-character "Tweets" every day.[2] Twitter's speed and ease of publication have made it an important communication medium for people from all walks of life. Twitter has played a prominent role in socio-political events, such as the Arab Spring[3] and the Occupy Wall Street movement.[4] Twitter has also been used to post damage reports and disaster preparedness information during large natural disasters, such as the Hurricane Sandy.

This book is for the reader who is interested in understanding the basics of collecting, storing, and analyzing Twitter data. The first half of this book discusses collection and storage of data. It starts by discussing how to collect Twitter data, looking at the free APIs provided by Twitter. We then goes on to discuss how to store this data for use in real-time applications. The second half is focused on analysis. Here, we focus on common measures and algorithms that are used to analyze social media data. We finish the analysis by discussing visual analytics, an approach which helps humans inspect the data through intuitive visualizations.

1.1 Main Takeaways from This Book

This book provides hands-on introduction to the collection and analysis of Twitter data. No knowledge of data analysis, or social network analysis is presumed. For all the concepts discussed in this book, we will provide in-depth description of the underlying assumptions and explain via construction of examples. The reader will

---

[1]http://twitter.com

[2]https://blog.twitter.com/2012/twitter-turns-six

[3]http://bit.ly/N6illb

[4]http://nyti.ms/SwZKVD

S. Kumar et al., *Twitter Data Analytics*, SpringerBriefs in Computer Science,
DOI 10.1007/978-1-4614-9372-3_1, © The Author(s) 2014

gain knowledge of the concepts in this book by building a crawler that collects Twitter data in real time. The reader will then learn how to analyze this data to find important time periods, users, and topics in their dataset. Finally, the reader will see how all of these concepts can be brought together to perform visual analysis and create meaningful software that uses Twitter data.

The code examples in this book are written in Java®, and JavaScript®. Familiarity with these languages will be useful in understanding the code, however the examples should be straightforward enough for anyone with basic programming experience. This book does assume that you know the programming concepts behind a high level language.

1.2 Learning Through Examples

Every concept discussed in this book is accompanied by illustrative examples. The examples in Chap. 4 use an open source network analysis library, JUNG™,[5] to perform network computations. The algorithms provided in this library are often highly optimized, and we recommend them for the development of production applications. However, because they are optimized, this code can be difficult to interpret for someone viewing these topics for the first time. In these cases, we present code that focuses more on readability than optimization to communicate the concepts using the examples. To build the visualizations in Chap. 5, we use the data visualization library D3™.[6] D3 is a versatile visualization toolkit, which supports various types of visualizations. We recommend the readers to browse through the examples to find other interesting ways to visualize Twitter data.

All of the examples read directly from a text file, where each line is a JSON document as returned by the Twitter APIs (the format of which is covered in Chap. 2). These examples can easily be manipulated to read from MongoDB®, but we leave this as an exercise for the reader.

Whenever "…" appears in a code example, code has been omitted from the example. This is done to remove code that is not pertinent to understanding the concepts. To obtain the full source code used in the examples, refer to the book's website, *http://tweettracker.fulton.asu.edu/tda*.

The dataset used for the examples in this book comes from the Occupy Wall Street movement, a protest centered around the wealth disparity in the US. This movement attracted significant focus on Twitter. We focus on a single day of this event to give a picture of what these measures look like with the same data. The dataset has been anonymized to remove any personally identifiable information. This dataset is also made available on the book's website for the reader to use when executing the examples.

---

[5]http://jung.sourceforge.net/

[6]http://d3js.org

To stay in agreement with Twitter's data sharing policies, some fields have been removed from this dataset, and others have been modified. When collecting data from the Twitter APIs in Chap. 2, you will get raw data with unaltered values for all of the fields.

1.3 Applying Twitter Data

Twitter's popularity as an information source has led to the development of applications and research in various domains. Humanitarian Assistance and Disaster Relief is one domain where information from Twitter is used to provide situational awareness to a crisis situation. Researchers have used Twitter to predict the occurrence of earthquakes [5] and identify relevant users to follow to obtain disaster related information [1]. Studies of Twitter's use in disasters include regions such as China [4], and Chile [2].

While a sampled view of Twitter is easily obtained through the APIs discussed in this book, the full view is difficult to obtain. The APIs only grant us access to a 1 % sample of the Twitter data, and concerns about the sampling strategy and the quality of Twitter data obtained via the API have been raised recently in [3]. This study indicates that care must be taken while constructing the queries used to collect data from the Streaming API.

References

1. S. Kumar, F. Morstatter, R. Zafarani, and H. Liu. Whom Should I Follow? Identifying Relevant Users During Crises. In *Proceedings of the 24th ACM conference on Hypertext and social media.* ACM, 2013.
2. M. Mendoza, B. Poblete, and C. Castillo. Twitter Under Crisis: Can we Trust What We RT? In *Proceedings of the First Workshop on Social Media Analytics*, 2010.
3. F. Morstatter, J. Pfeffer, H. Liu, and K. Carley. Is the Sample Good Enough? Comparing Data from Twitter's Streaming API with Twitter's Firehose. In *International AAAI Conference on Weblogs and Social Media*, 2013.
4. Y. Qu, C. Huang, P. Zhang, and J. Zhang. Microblogging After a Major Disaster in China: A Case Study of the 2010 Yushu Earthquake. In *Computer Supported Cooperative Work and Social Computing*, pages 25–34, 2011.
5. T. Sakaki, M. Okazaki, and Y. Matsuo. Earthquake Shakes Twitter Users: Real-Time Event Detection by Social Sensors. In *Proceedings of the 19th international conference on World wide web*, pages 851–860. ACM, 2010.

Chapter 2
Crawling Twitter Data

Users on Twitter generate over 400 million Tweets everyday.[1] Some of these Tweets are available to researchers and practitioners through public APIs at no cost. In this chapter we will learn how to extract the following types of information from Twitter:

- Information about a user,
- A user's network consisting of his connections,
- Tweets published by a user, and
- Search results on Twitter.

APIs to access Twitter data can be classified into two types based on their design and access method:

- REST APIs are based on the REST architecture[2] now popularly used for designing web APIs. These APIs use the pull strategy for data retrieval. To collect information a user must explicitly request it.
- Streaming APIs provides a continuous stream of public information from Twitter. These APIs use the push strategy for data retrieval. Once a request for information is made, the Streaming APIs provide a continuous stream of updates with no further input from the user.

They have different capabilities and limitations with respect to what and how much information can be retrieved. The Streaming API has three types of endpoints:

- Public streams: These are streams containing the public Tweets on Twitter.
- User streams: These are single-user streams, with to all the Tweets of a user.
- Site streams: These are multi-user streams and intended for applications which access Tweets from multiple users.

---

[1]http://articles.washingtonpost.com/2013-03-21/business/37889387_1_tweets-jack-dorsey-twitter

[2]http://en.wikipedia.org/wiki/Representational_state_transfer

S. Kumar et al., *Twitter Data Analytics*, SpringerBriefs in Computer Science, DOI 10.1007/978-1-4614-9372-3_2, © The Author(s) 2014

As the Public streams API is the most versatile Streaming API, we will use it in all the examples pertaining to Streaming API.

In this chapter, we illustrate how the aforementioned types of information can be collected using both forms of Twitter API. Requests to the APIs contain parameters which can include hashtags, keywords, geographic regions, and Twitter user IDs. We will explain the use of parameters in greater detail in the context of specific APIs later in the chapter. Responses from Twitter APIs is in JavaScript Object Notation (JSON) format.[3] JSON is a popular format that is widely used as an object notation on the web.

Twitter APIs can be accessed only via authenticated requests. Twitter uses *Open Authentication* and each request must be signed with valid Twitter user credentials. Access to Twitter APIs is also limited to a specific number of requests within a time window called the *rate limit*. These limits are applied both at individual user level as well as at the application level. A *rate limit window* is used to renew the quota of permitted API calls periodically. The size of this window is currently 15 min.

We begin our discussion with a brief introduction to OAuth.

2.1 Introduction to Open Authentication (OAuth)

Open Authentication (OAuth) is an open standard for authentication, adopted by Twitter to provide access to protected information. Passwords are highly vulnerable to theft and OAuth provides a safer alternative to traditional authentication approaches using a three-way handshake. It also improves the confidence of the user in the application as the user's password for his Twitter account is never shared with third-party applications.

The authentication of API requests on Twitter is carried out using OAuth. Figure 2.1 summarizes the steps involved in using OAuth to access Twitter API. Twitter APIs can only be accessed by applications. Below we detail the steps for making an API call from a Twitter application using OAuth:

1. Applications are also known as consumers and all applications are required to register themselves with Twitter.[4] Through this process the application is issued a consumer key and secret which the application must use to authenticate itself to Twitter.
2. The application uses the consumer key and secret to create a unique Twitter link to which a user is directed for authentication. The user authorizes the application by authenticating himself to Twitter. Twitter verifies the user's identity and issues a OAuth verifier also called a PIN.

---

[3]http://en.wikipedia.org/wiki/JSON

[4]Create your own application at http://dev.twitter.com

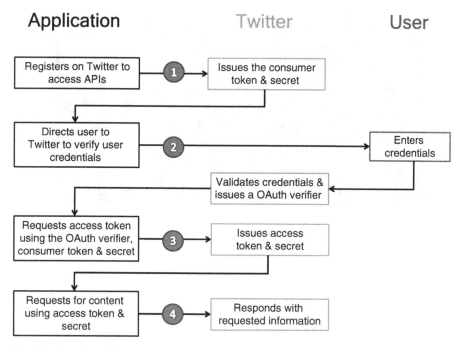

Fig. 2.1 OAuth workflow

3. The user provides this PIN to the application. The application uses the PIN to request an "Access Token" and "Access Secret" unique to the user.
4. Using the "Access Token" and "Access Secret", the application authenticates the user on Twitter and issues API calls on behalf of the user.

The "Access Token" and "Access Secret" for a user do not change and can be cached by the application for future requests. Thus, this process only needs to be performed once, and it can be easily accomplished using the method *GetUserAccessKeySecret* in Listing 2.1.

2.2 Collecting a User's Information

On Twitter, users create profiles to describe themselves to other users on Twitter. A user's profile is a rich source of information about him. An example of a Twitter user's profile is presented in Fig. 2.2. Following distinct pieces of information regarding a user's Twitter profile can be observed in the figure:

Fig. 2.2 An example of a Twitter profile

Listing 2.1 Generating OAuth token for a user

```
public OAuthTokenSecret GetUserAccessKeySecret() {
        . . .
        //Step 1 is performed directly on twitter.com after
            registration.
        //Step 2 User authenticates on twitter.com and generates
            a PIN
        OAuthConsumer consumer = new CommonsHttpOAuthConsumer(
            OAuthUtils.CONSUMER_KEY, OAuthUtils.
            CONSUMER_SECRET);
        OAuthProvider provider = new DefaultOAuthProvider(
            OAuthUtils.REQUEST_TOKEN_URL, OAuthUtils.
            ACCESS_TOKEN_URL, OAuthUtils.AUTHORIZE_URL);
        String authUrl = provider.retrieveRequestToken(consumer,
            OAuth.OUT_OF_BAND);
        //Visit authUrl and enter the PIN in the application
        BufferedReader br = new BufferedReader(new
            InputStreamReader(System.in));
        String pin = br.readLine();
        //Step 3 Twitter generates the token and secret using
            the provided PIN
        provider.retrieveAccessToken(consumer,pin);
        String accesstoken = consumer.getToken();
        String accesssecret  = consumer.getTokenSecret();
        OAuthTokenSecret tokensecret = new OAuthTokenSecret(
            accesstoken,accesssecret);
        return tokensecret;
        . . .
}
Source: Chapter2/openauthentication/OAuthExample.java
```

- User's real name (Data Analytics)
- User's Twitter handle(@twtanalyticsbk)
- User's location (Tempe, AZ)
- URL, which typically points to a more detailed profile of the user on an external website (tweettracker.fulton.asu.edu/tda)
- Textual description of the user and his interests (Twitter Data Analytics is a book for...)
- User's network activity information on Twitter (1 follower and following 6 friends)
- Number of Tweets published by the user (1 Tweet)
- Verified mark if the identity of the user has been externally verified by Twitter
- Profile creation date

Listing 2.2 Using Twitter API to fetch a user's profile

```java
public JSONObject GetProfile(String username) {
    . . .
    // Step 1: Create the API request using the supplied
        username
    URL url = new URL("https://api.twitter.com/1.1/users/
        show.json?screen_name="+username);
    HttpURLConnection huc = (HttpURLConnection) url.
        openConnection();
    huc.setReadTimeout(5000);
    // Step 2: Sign the request using the OAuth Secret
    consumer.sign(huc);
    huc.connect();
    . . .
    /** Step 3: If the requests have been exhausted,
    * then wait until the quota is renewed
    */
    if(huc.getResponseCode()==429) {
            try {
                    huc.disconnect();
                    Thread.sleep(this.GetWaitTime("/users/
                        show/:id"));
                    flag = false;
    . . .
    // Step 4: Retrieve the user's profile from Twitter
    bRead = new BufferedReader(new InputStreamReader((
        InputStream) huc.getContent()));
    . . .
    profile = new JSONObject(content.toString());
    . . .
    return userobj;
}
```

Source: Chapter2/restapi/RESTApiExample.java

Listing 2.3 A sample Twitter user object

```
{
    "location": "Tempe,AZ",
    "default_profile": true,
    "statuses_count": 1,
    "description": "Twitter Data Analytics is a book for
        practitioners and researchers interested in
        investigating Twitter data.",
    "verified": false,
    "name": "DataAnalytics",
    "created_at": "Tue Mar 12 18:43:47 +0000 2013",
    "followers_count": 1,
    "geo_enabled": false,
    "url": "http://t.co/HnlG9amZzj",
    "time_zone": "Arizona",
    "friends_count": 6,
    "screen_name": "twtanalyticsbk",
    //Other user fields
         . . .
}
```

Using the API *users/show*,[5] a user's profile information can be retrieved using the method *GetProfile*. The method is presented in Listing 2.2. It accepts a valid username as a parameter and fetches the user's Twitter profile.

Key Parameters: Each user on Twitter is associated with a unique id and a unique Twitter handle which can be used to retrieve his profile. A user's Twitter handle, also called their screen name (screen_name), or the Twitter ID of the user (user_id), is mandatory. A typical user object is formatted as in Listing 2.3.

Rate Limit: A maximum of 180 API calls per single user and 180 API calls from a single application are accepted within a single rate limit window.

Note: User information is generally included when Tweets are fetched from Twitter. Although the Streaming API does not have a specific endpoint to retrieve user profile information, it can be obtained from the Tweets fetched using the API.

2.3   Collecting a User's Network

A user's network consists of his connections on Twitter. Twitter is a directed network and there are two types of connections between users. In Fig. 2.3, we can observe an example of the nature of these edges. John follows Alice, therefore John is Alice's follower. Alice follows Peter, hence Peter is a friend of Alice.

---

[5]https://dev.twitter.com/docs/api/1.1/get/users/show

Fig. 2.3 An example of a
Twitter network with different
types of edges

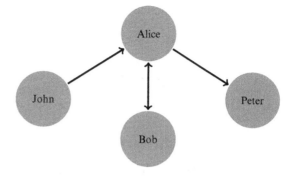

Listing 2.4 Using the Twitter API to fetch the followers of a user

```
public JSONArray GetFollowers(String username) {
        . . .
        // Step 1: Create the API request using the supplied
           username
        URL url = new URL("https://api.twitter.com/1.1/followers
           /list.json?screen_name="+username+"&cursor="
           + cursor);
        HttpURLConnection huc = (HttpURLConnection) url.
           openConnection();
        huc.setReadTimeout(5000);
        // Step 2: Sign the request using the OAuth Secret
        Consumer.sign(huc);
        huc.connect();
        . . .
        /** Step 3: If the requests have been exhausted,
          * then wait until the quota is renewed
          */
        if(huc.getResponseCode()==429) {
                try {
                        Thread.sleep(this.GetWaitTime("/
                           followers/list"));
                } catch (InterruptedException ex) {
                        Logger.getLogger(RESTApiExample.class.
                           getName()).log(Level.SEVERE, null,
                           ex);
                }
        }
        // Step 4: Retrieve the followers list from Twitter
        bRead = new BufferedReader(new InputStreamReader((
           InputStream) huc.getContent()));
        StringBuilder content = new StringBuilder();
        String temp = "";
        while((temp = bRead.readLine())!=null) {
                content.append(temp);
        }
        try {
```

```
        JSONObject jobj = new JSONObject(content.
            toString());
        // Step 5: Retrieve the token for the next
            request
        cursor = jobj.getLong("next_cursor");
        JSONArray idlist = jobj.getJSONArray("users");
        for(int i=0;i<idlist.length();i++) {
                followers.put(idlist.getJSONObject(i));
        }
     . . .
     return followers;
}
```
Source: Chapter2/restapi/RESTApiExample.java

2.3.1 Collecting the Followers of a User

The followers of a user can be crawled from Twitter using the endpoint *followers/list*,[6] by employing the method *GetFollowers* summarized in Listing 2.4. The response from Twitter consists of an array of user profile objects such as the one described in Listing 2.3

Key Parameters: screen_name or user_id is mandatory to access the API. Each request returns a maximum of 15 followers of the specified user in the form of a Twitter User object. The parameter "cursor" can be used to paginate through the results. Each request returns the cursor for use in the request for the next page.

Rate Limit: A maximum of 15 API calls from a user and 30 API calls from an application are allowed within a rate limit window.

2.3.2 Collecting the Friends of a User

The friends of a user can be crawled using the Twitter API *friends/list*[7] by employing the method *GetFriends*, which is summarized in Listing 2.5. The method constructs a call to the API and takes a valid Twitter username as the parameter. It uses the cursor to retrieve all the friends of a user and if the API limit is reached, it will wait until the quota has been renewed.

Key Parameters: As with the followers API, a valid screen_name or user_id is mandatory. Each request returns a list of 20 friends of a user as Twitter User objects. The parameter "cursor" can be used to paginate through the results. Each request returns the cursor to be used in the request for the next page.

---

[6]https://dev.twitter.com/docs/api/1.1/get/followers/list

[7]https://dev.twitter.com/docs/api/1.1/get/friends/list

Listing 2.5 Using the Twitter API to fetch the friends of a user

```java
public JSONArray GetFriends(String username) {
        . . .
        JSONArray friends = new JSONArray();
        // Step 1: Create the API request using the supplied
            username
        URL url = new URL("https://api.twitter.com/1.1/friends/
            list.json?screen_name="+username+"&cursor="+cursor);
        HttpURLConnection huc = (HttpURLConnection) url.
            openConnection();
        huc.setReadTimeout(5000);
        // Step 2: Sign the request using the OAuth Secret
        Consumer.sign(huc);
        huc.connect();
        . . .
        /** Step 3: If the requests have been exhausted,
         * then wait until the quota is renewed
         */
        if(huc.getResponseCode()==429) {
                try {
                        Thread.sleep(this.GetWaitTime("/friends/
                            list"));
                } catch (InterruptedException ex) {
                        Logger.getLogger(RESTApiExample.class.
                            getName()).log(Level.SEVERE, null,
                            ex);
                }
        }
        // Step 4: Retrieve the friends list from Twitter
        bRead = new BufferedReader(new InputStreamReader((
            InputStream) huc.getContent()));
        . . .
        JSONObject jobj = new JSONObject(content.toString());
        // Step 5: Retrieve the token for the next request
        cursor = jobj.getLong("next_cursor");
        JSONArray userlist = jobj.getJSONArray("users");
        for(int i=0;i<userlist.length();i++) {
                friends.put(userlist.get(i));
        }
        . . .
        return friends;
}
```

Source: Chapter2/restapi/RESTApiExample.java

Rate Limit: A maximum of 15 calls from a user and 30 API calls from an application are allowed within a rate limit window.

2.4 Collecting a User's Tweets

A Twitter user's Tweets are also known as status messages. A Tweet can be at most 140 characters in length. Tweets can be published using a wide range of mobile and desktop clients and through the use of Twitter API. A special kind of Tweet is the retweet, which is created when one user reposts the Tweet of another user. We will discuss the utility of retweets in greater detail in Chaps. 4 and 5.

A user's Tweets can be retrieved using both the REST and the Streaming API.

2.4.1 REST API

We can access a user's Tweets by using *statuses/user_timeline*[8] from the REST APIs. Using this API, one can retrieve 3,200 of the most recent Tweets published by a user including retweets. The API returns Twitter "Tweet" objects shown in Listing 2.6.

An example describing the process to access this API can be found in the *GetStatuses* method summarized in Listing 2.7.

Key Parameters: We can retrieve 200 Tweets on each page we collect. The parameter max_id is used to paginate through the Tweets of a user. To retrieve the next page we use the ID of the oldest Tweet in the list as the value of this parameter in the subsequent request. Then, the API will retrieve only those Tweets whose IDs are below the supplied value.

Rate Limit: An application is allowed 300 requests within a rate limit window and up to 180 requests can be made using the credentials of a user.

Listing 2.6 An example of Twitter Tweet object

```
{
    "text": "This is the first tweet.",
    "lang": "en",
    "id": 352914247774248960,
    "source": "web",
    "retweet_count": 0,
    "created_at": "Thu Jul 04 22:18:08 +0000 2013",
    //Other Tweet fields
    . . .
    "place": {
        "place_type": "city",
        "name": "Tempe",
        "country_code": "US",
        "url": "https://api.twitter.com/1.1/geo/id/7
            cb7440bcf83d464.json",
        "country": "United States",
```

---

[8]https://dev.twitter.com/docs/api/1.1/get/statuses/user_timeline

```
        "full_name": "Tempe, AZ",
        //Other place fields
           . . .
    },
    "user": {
        //User Information in the form of Twitter user object
           . . .
    }
}
```

Listing 2.7 Using the Twitter API to fetch the Tweets of a user

```
public JSONArray GetStatuses(String username) {
        . . .
        // Step 1: Create the API request using the supplied
            username
        // Use (max_id-1) to avoid getting redundant Tweets.
        url = new URL("https://api.twitter.com/1.1/statuses/
            user_timeline.json?screen_name=" + username+"&
            include_rts="+include_rts+"&count="+tweetcount+"&
            max_id="+(maxid-1));
        HttpURLConnection huc = (HttpURLConnection) url.
            openConnection();
        huc.setReadTimeout(5000);
        // Step 2: Sign the request using the OAuth Secret
        Consumer.sign(huc);
        /** Step 3: If the requests have been exhausted,
          * then wait until the quota is renewed  */
        . . .
        //Step 4: Retrieve the Tweets from Twitter
        bRead = new BufferedReader(new InputStreamReader((
            InputStream) huc.getInputStream()));
        . . .
        for(int i=0;i<statusarr.length();i++) {
                JSONObject jobj = statusarr.getJSONObject(i);
                statuses.put(jobj);
                // Step 5: Get the id of the oldest Tweet ID as
                    max_id to retrieve the next batch of Tweets
                if(!jobj.isNull("id")) {
                        maxid = jobj.getLong("id");
        . . .
    return statuses;
}
```
Source: Chapter2/restapi/RESTApiExample.java

2.4.2   Streaming API

Specifically, the *statuses/filter*[9] API provides a constant stream of public Tweets published by a user. Using the method *CreateStreamingConnection* summarized in Listing 2.8, we can create a POST request to the API and fetch the search results as a stream. The parameters are added to the request by reading through a list of userids using the method *CreateRequestBody*, which is summarized in Listing 2.9.

Listing 2.8   Using the Streaming API to fetch Tweets

```
public void CreateStreamingConnection(String baseUrl, String
    outFilePath) {
        HttpClient httpClient = new DefaultHttpClient();
        httpClient.getParams().setParameter(CoreConnectionPNames
            .CONNECTION_TIMEOUT, new Integer(90000));
        //Step 1: Initialize OAuth Consumer
        OAuthConsumer consumer = new CommonsHttpOAuthConsumer(
            OAuthUtils.CONSUMER_KEY,OAuthUtils.CONSUMER_SECRET);
        consumer.setTokenWithSecret(OAuthToken.getAccessToken(),
            OAuthToken.getAccessSecret());
        //Step 2: Create a new HTTP POST request and set
            parameters
        HttpPost httppost = new HttpPost(baseUrl);
        try {
            httppost.setEntity(new UrlEncodedFormEntity(
                CreateRequestBody(), "UTF-8"));
            . . .
            //Step 3: Sign the request
            consumer.sign(httppost);
            . . .
            HttpResponse response;
            InputStream is = null;
            try {
                //Step 4: Connect to the API
                response = httpClient.execute(httppost);
                . . .
                HttpEntity entity = response.getEntity();
                try {
            is = entity.getContent();
                    . . .
                    //Step 5: Process the incoming Tweet
                        Stream
                    this.ProcessTwitterStream(is, outFilePath
                        );
                    . . .
}
```
Source: Chapter2/streamingapi/StreamingApiExample.java

---

[9]https://dev.twitter.com/docs/api/1.1/post/statuses/filter

Listing 2.9 Adding parameters to the Streaming API

```java
private List<NameValuePair> CreateRequestBody() {
        List<NameValuePair> params = new ArrayList<NameValuePair
            >();
        if(Userids != null&&Userids.size()>0) {
                //Add userids
                params.add(CreateNameValuePair("follow",
                    Userids));
        }
        if (Geoboxes != null&&Geoboxes.size()>0) {
                //Add geographic bounding boxes
                params.add(CreateNameValuePair("locations",
                    Geoboxes));
        }
        if (Keywords != null&&Keywords.size()>0) {
                //Add keywords/hashtags/phrases
                params.add(CreateNameValuePair("track",
                    Keywords));
        }
        return params;
}
```
Source: Chapter2/streamingapi/StreamingApiExample.java

Key Parameters: The `follow`[10] parameter can be used to specify the userids of 5,000 users as a comma separated list.

Rate Limit: Rate limiting works differently in the Streaming API. In each connection an application is allowed to submit up to 5,000 Twitter userids. Only public Tweets published by the user can be captured using this API.

2.5 Collecting Search Results

Search on Twitter is facilitated through the use of parameters. Acceptable parameter values for search include keywords, hashtags, phrases, geographic regions, and usernames or userids. Twitter search is quite powerful and is accessible by both the REST and the Streaming APIs. There are certain subtle differences when using each API to retrieve search results.

2.5.1 REST API

Twitter provides the *search/tweets* API to facilitate searching the Tweets. The search API takes words as queries and multiple queries can be combined as a comma separated list. Tweets from the previous 10 days can be searched using this API.

---

[10]https://dev.twitter.com/docs/streaming-apis/parameters#follow

Listing 2.10 Searching for Tweets using the REST API

```java
public JSONArray GetSearchResults(String query) {
    try {
        // Step 1:
        String URL_PARAM_SEPERATOR = "&";
        StringBuilder url = new StringBuilder();
        url.append("https://api.twitter.com/1.1/search/tweets.
            json?q=");
        //query needs to be encoded
        url.append(URLEncoder.encode(query, "UTF-8"));
        url.append(URL_PARAM_SEPERATOR);
        url.append("count=100");
        URL navurl = new URL(url.toString());
        HttpURLConnection huc = (HttpURLConnection) navurl.
            openConnection();
        huc.setReadTimeout(5000);
        Consumer.sign(huc);
        huc.connect();
        . . .
        // Step 2: Read the retrieved search results
        BufferedReader bRead = new BufferedReader(new
            InputStreamReader((InputStream) huc.getInputStream()
            ));
        String temp;
        StringBuilder page = new StringBuilder();
        while( (temp = bRead.readLine())!=null) {
                page.append(temp);
        }
        JSONTokener jsonTokener = new JSONTokener(page.toString
            ());
        try{
            JSONObject json = new JSONObject(jsonTokener);
            //Step 4: Extract the Tweet objects as an array
            JSONArray results = json.getJSONArray("statuses");
            return results;
            . . .
}
```

Source: Chapter2/restapi/RESTApiExample.java

Requests to the API can be made using the method *GetSearchResults* presented in Listing 2.10. Input to the function is a keyword or a list of keywords in the form of an OR query. The function returns an array of Tweet objects.

Key Parameters: result_type parameter can be used to select between the top ranked Tweets, the latest Tweets, or a combination of the two types of search results matching the query. The parameters max_id and since_id can be used to paginate through the results, as in the previous API discussions.

Rate Limit: An application can make a total of 450 requests and up to 180 requests from a single authenticated user within a rate limit window.

2.5.2 Streaming API

Using the Streaming API, we can search for keywords, hashtags, userids, and geographic bounding boxes simultaneously. The *filter* API facilitates this search and provides a continuous stream of Tweets matching the search criteria. POST method is preferred while creating this request because when using the GET method to retrieve the results, long URLs might be truncated. Listings 2.8 and 2.9 describe how to connect to the Streaming API with the supplied parameters.

Listing 2.11 Processing the streaming search results

```java
public void ProcessTwitterStream(InputStream is, String
    outFilePath) {
    BufferedWriter bwrite = null;
    try {
      /** A connection to the streaming API is already
       *  created and the response is  contained in
       *  the InpuStream
       */
        JSONTokener jsonTokener = new JSONTokener(new
            InputStreamReader(is, ''UTF-8''));
        ArrayList<JSONObject> rawtweets = new ArrayList<
            JSONObject>();
        int nooftweetsuploaded = 0;
      //Step 1: Read until the stream is exhausted
        while(true) {
        try {
            JSONObject temp = new JSONObject(jsonTokener);
            rawtweets.add(temp);
            if (rawtweets.size() >= RECORDS_TO_PROCESS){
            Calendar cal = Calendar.getInstance();
            String filename = outFilePath + ''tweets_'' +
                cal.getTimeInMillis() + ''.json'';
                //Step 2: Periodically write the
                    processed Tweets to a file
                bwrite = new BufferedWriter(new
                    OutputStreamWriter(new
                    FileOutputStream(filename),
                    ''UTF-8''));
                nooftweetsuploaded+=RECORDS_TO_PROCESS;
                for (JSONObject jobj : rawtweets) {
            bwrite.write(jobj.toString());
                bwrite.newLine();
                }
                bwrite.close();
                rawtweets.clear();
                . . .
}
```
Source: Chapter2/streamingapi/StreamingApiExample.java

In method *ProcessTwitterStream*, as in Listing 2.11, we show how the incoming stream is processed. The input is read in the form of a continuous stream and each Tweet is written to a file periodically. This behavior can be modified as per the requirement of the application, such as storing and indexing the Tweets in a database. More discussion on the storage and indexing of Tweets will follow in Chap. 3.

Key Parameters: There are three key parameters:

- `follow`: a comma-separated list of userids to follow. Twitter returns all of their public Tweets in the stream.
- `track`: a comma-separated list of keywords to track. Multiple keywords are provided as a comma separated list.
- `locations`: a comma-separated list of geographic bounding box containing the coordinates of the southwest point and the northeast point as (longitude, latitude) pairs.

Rate Limit: Streaming APIs limit the number of parameters which can be supplied in one request. Up to 400 keywords, 25 geographic bounding boxes and 5,000 userids can be provided in one request. In addition, the API returns all matching documents up to a volume equal to the streaming cap. This cap is currently set to 1% of the total current volume of Tweets published on Twitter.

2.6 Strategies to Identify the Location of a Tweet

Location information on Twitter is available from two different sources:

- Geotagging information: Users can optionally choose to provide location information for the Tweets they publish. This information can be highly accurate if the Tweet was published using a smartphone with GPS capabilities.
- Profile of the user: User location can be extracted from the location field in the user's profile. The information in the location field itself can be extracted using the APIs discussed above.

Approximately 1% of all Tweets published on Twitter are geolocated. This is a very small portion of the Tweets, and it is often necessary to use the profile information to determine the Tweet's location. This information can be used in different visualizations as you will see in Chap. 5. The location string obtained from the user's profile must first be translated into geographic coordinates. Typically, a gazetteer is used to perform this task. A gazetteer takes a location string as input, and returns the coordinates of the location that best correspond to the string. The granularity of the location is generally coarse. For example, in the case of large regions, such as cities, this is usually the center of the city. There are several online gazetteers which provide this service, including Bing™, Google™, and MapQuest™. In our example, we will use the Nominatim service from MapQuest[11]

---

[11]http://developer.mapquest.com/web/products/open/nominatim

Listing 2.12 Translating location string into coordinates

```java
public Location TranslateLoc(String loc) {
    if(loc!=null&&!loc.isEmpty()) {
        String encodedLoc="";
        try {
        // Step 1: Encode the location name
                encodedLoc = URLEncoder.encode(loc, "UTF-8");
        . . .
        /** Step 2: Create a get request to MapQuest API with
            the
         * name of the location
         */
        String url= "http://open.mapquestapi.com/nominatim/v1/
            search?q="+encodedLoc+"&format=json";
        String page = ReadHTML(url);
        if(page!=null) {
                try{
                        JSONArray results = new JSONArray(page);
                        if(results.length()>0) {
                                //Step 3: Read and extract the
                                    coordinates of the location
                                    as a JSONObject
                                Location loca = new Location(
                                    results.getJSONObject(0).
                                    getDouble("lat"),results.
                                    getJSONObject(0).getDouble("
                                    lon"));
                                return loca;
        . . .
}
```

Source: Chapter2/location/LocationTranslationExample.java

to demonstrate this process. In Listing 2.12, a summary of the method *TranslateLoc* is provided, which is defined in the class *LocationTranslateExample*. The response is provided in JSON, from which the coordinates can be easily extracted. If the service is unable to find a match, it will return (0,0) as the coordinates.

2.7 Obtaining Data via Resellers

The rate limitations of Twitter APIs can be too restrictive for certain types of applications. To satisfy such requirements, Twitter Firehose provides access to 100% of the public Tweets on Twitter at a price. Firehose data can be purchased through third party resellers of Twitter data. At the time of writing of this book, there are three resellers of data, each of which provide different levels of access. In addition to Twitter data some of them also provide data from other social media platforms, which might be useful while building social media based systems. These include the following:

- DataSift™[12] – provides access to past data as well as streaming data
- GNIP™[13] – provides access to streaming data only
- Topsy™[14] – provides access to past data only

2.8   Further Reading

Full documentation of v1.1 of the Twitter API can be found at [1]. It also contains the most up-to-date and detailed information on the rate limits applicable to individual APIs. Twitter HTTP Error Codes & Responses [2] contains a list of HTTP error codes returned by the Twitter APIs. It is a useful resource while debugging applications. The REST API for search accepts several different parameters to facilitate the construction of complex queries. A full list of these along with examples can be found in [4]. The article further clarifies on what is possible using the Search API and explains the best practices for accessing the API. Various libraries exist in most popular programming languages, which encapsulate the complexity of accessing the Twitter API by providing convenient methods. A full list of all available libraries can be found in [3]. Twitter has also released an open source library of their own called the Hosebird, which has been tested to handle firehose streams.

References

1. Twitter. Twitter API v1.1 Documentation. https://dev.twitter.com/docs/api/1.1, 2013. [Online; accessed 19-March-2013].
2. Twitter. Twitter HTTP Error Codes & Responses. https://dev.twitter.com/docs/error-codes-responses, 2013. [Online; accessed 19-March-2013].
3. Twitter. Twitter Libraries. https://dev.twitter.com/docs/twitter-libraries, 2013. [Online; accessed 9-July-2013].
4. Twitter. Using the Twitter Search API. https://dev.twitter.com/docs/using-search, 2013. [Online; accessed 9-July-2013].

---

[12]http://datasift.com

[13]http://gnip.com

[14]http://topsy.com

Chapter 3
Storing Twitter Data

In the previous chapter, we covered data collection methodologies. Using these methods, one can quickly amass a large volume of Tweets, Tweeters, and network information. Managing even a moderately-sized dataset is cumbersome when storing data in a text-based archive, and this solution will not give the performance needed for a real-time application. In this chapter we present some common storage methodologies for Twitter data using NoSQL.

3.1 NoSQL Through the Lens of MongoDB

Keeping track of every purchase, click, and "like" has caused the data needs of many companies to double every 14 months. There has been an explosion in the size of data generated on social media. This data explosion calls for a new data storage paradigm. At the forefront of this movement is NoSQL [3], which promises to store big data in a more accessible way than the traditional, relational model.

There are several NoSQL implementations. In this book, we choose MongoDB[1] as an example NoSQL implementation. We choose it for its adherence to the following principles:

- **Document-Oriented Storage.** MongoDB stores its data in JSON-style objects. This makes it very easy to store raw documents from Twitter's APIs.
- **Index Support.** MongoDB allows for indexes on any field, which makes it easy to create indexes optimized for your application.
- **Straightforward Queries.** MongoDB's queries, while syntactically much different from SQL, are semantically very similar. In addition, MongoDB supports MapReduce, which allows for easy lookups in the data.

---

[1]http://www.mongodb.org/

S. Kumar et al., *Twitter Data Analytics*, SpringerBriefs in Computer Science,
DOI 10.1007/978-1-4614-9372-3_3, © The Author(s) 2014

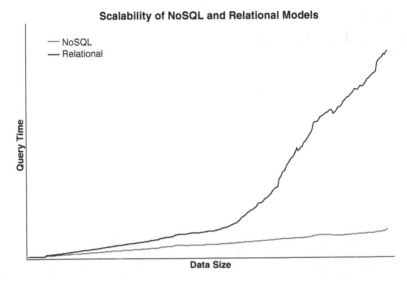

Fig. 3.1 Comparison of traditional relational model with NoSQL model. As data grows to a large capacity, the NoSQL database outpaces the relational model

- **Speed.** Figure 3.1 shows a comparison of query speed between the relational model and MongoDB.

In addition to these abilities, it also works well in a single-instance environment, making it easy to set up on a home computer and run the examples in this chapter.

3.2 Setting Up MongoDB on a Single Node

The most simple configuration of MongoDB is a single instance running on one machine. This setup allows for access to all of the features of MongoDB. We use MongoDB 2.4.4,[2] the latest version at the time of this writing.

3.2.1 Installing MongoDB on Windows®

1. Obtain the latest version of MongoDB from http://www.mongodb.org/ downloads. Extract the downloaded zip file.
2. Rename the extracted folder to mongodb.
3. Create a folder called data next to the mongodb folder.

---

[2]http://docs.mongodb.org/manual/

4. Create a folder called db within the data folder. Your file structure should reflect that shown below.

3.2.2 Running MongoDB on Windows

1. Open the command prompt and move to the directory above the mongodb folder.
2. Run the command mongodb\bin\mongod.exe-dbpath data\db
3. If Windows prompts you, make sure to allow MongoDB to communicate on private networks, but not public ones. Without special precautions, MongoDB should not be run in an open environment.
4. Open another command window and move to the directory where you put the mongodb folder.
5. Run the command mongodb\bin\mongo.exe. This is the command-line interface to MongoDB. You can now issue commands to MongoDB.

3.2.3 Installing MongoDB on Mac OS X®

1. Obtain the latest version of MongoDB from http://www.mongodb.org/downloads.
2. Rename the downloaded file to mongodb.tgz.
3. Open the "Terminal" application. Move to the folder where you downloaded MongoDB.
4. Run the command tar -zxvf mongodb.tgz. This will create a folder with the name mongodb-osx-[platform]-[version] in the same directory. For version 2.4.4, this folder will be called mongodb-osx-x86_64-2.4.4.
5. Run the command mv -n mongodb-osx-[platform]-[version]/mongodb. This will give us a more convenient folder name.
6. Run the command mkdir data && mkdir data/db. This will create the subfolders where we will store our data.

3.2.4   *Running MongoDB on Mac OS X*

1. Open the "Terminal" application and move to the directory above the `mongodb` folder.
2. Run the command `./mongodb/bin/mongod-dbpath data/db`
3. Open another tab in Terminal (Cmd-T).
4. Run the command `./mongodb/bin/mongo`. This is the command-line interface to MongoDB. You can now issue commands to MongoDB.

3.3   MongoDB's Data Organization

MongoDB organizes its data in the following hierarchy: database, collection, document. A database is a set of collections, and a collection is a set of documents. The organization of data in MongoDB is shown in Fig. 3.2. Here we will demonstrate how to interact with each level in this hierarchy to store data.

3.4   How to Execute the MongoDB Examples

The examples presented in this chapter are written in JavaScript – the language underpinning MongoDB. To run these examples, do the following:

1. Run `mongod`, as shown above. The process doing this varies is outlined in Sect. 3.2.

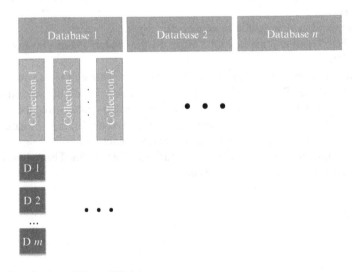

Fig. 3.2 Organization of MongoDB data

2. Change directories to your `bin` folder: `cd mongodb/bin`.
3. Execute the following command: `mongo localhost/tweetdata path/ to/example.js`. This will run the example on your local MongoDB installation. If you are on windows, you will have to replace `mongo` with `mongo.exe`.

3.5   Adding Tweets to the Collection

Now that we have a collection in the database, we will add some Tweets to it. Because MongoDB uses JSON to store its documents, we can import the data *exactly* as it was collected from Twitter, with no need to map columns. To load this, download the Occupy Wall Street data included in the supplementary materials, `ows.json`. Next, with mongod running, issue the following command[3]:

`mongoimport -d tweetdata -c tweets -file ows.json`

`mongoimport` is a utility that is packaged with MongoDB that allows you to import JSON documents. By running the above command, we have added all of the JSON documents in the file to the collection we created earlier. We now have some Tweets stored in our database, and we are ready to issue some commands to analyze this data.

3.6   Optimizing Collections for Queries

To make our documents more accessible, we will extract some key features for indexing later. For example, while the "created_at" field gives us information about a date in a readable format, converting it to a JavaScript date each time we do a date comparison will add overhead to our computations. It makes sense to add a field "timestamp" whose value contains the Unix timestamp[4] representing the information contained in "created_at". This redundancy trades disk space for efficient computation, which is more of a concern when building real-time applications which rely on big data. Listing 3.1 is a post-processing script that adds fields that make handling the Twitter data more convenient and efficient.

---

[3]On Windows, you exchange `mongoimport` with `mongoimport.exe`.
[4]A number, the count of milliseconds since January 1st, 1970.

Listing 3.1 Post-processing step to add extra information to data

```
> //enumerate each Tweet
> db.tweets.find().forEach(function(doc){
...     //save the time string in Unix time.
...     doc.timestamp = +new Date(doc.created_at);
...     //reduce the geobox to one point
...     doc.geoflag = !!doc.coordinates;
...     if(doc.coordinates && doc.coordinates.coordinates){
...         doc.location = {"lat": doc.coordinates.coordinates
    [1], "lng": doc.coordinates.coordinates[0]};
...     }
...     //save a lowercased version of the screen name
...     doc.screen_name_lower = doc.user.screen_name.toLowerCase
    ();
...     //save our modifications
...     db.tweets.save(doc);
... });
Source: Chapter3/postProcessingExample.js
```

Listing 3.2 Create an index on the "timestamp" field

```
> db.tweets.ensureIndex({"timestamp": 1})
```

3.7 Indexes

We now have inserted some documents into a collection, but as they stand querying them will be slow as we have not created any indexes. That is, we have not told MongoDB which fields in the document to optimize for faster lookup.

One of the most important concepts to understand for fast access of a MongoDB collection is indexing. The indexes you choose will depend largely on the queries that you run often, those that *must* be executed in real time. While the indexes you choose will depend on your data, here we will show some indexes that are often useful in querying Twitter data in real-time.

The first index we create will be on our "timestamp" field. This command is shown in Listing 3.2.

When creating an index, there are several rules MongoDB enforces to ensure that an index is used:

- **Only one index is used per query.** While you can create as many indexes as you want for a given collection, you can only use one for each query. If you have multiple fields in your query, you can create a "compound index" on both fields. For example, if you want to create an index on "timestamp", and then "retweet_count", can pass {"timestamp": 1, "retweet_count": 1}.

- **Indexes can only use fields in the order they were created.** Say, for example, we create the index `{"timestamp": 1, "retweet_count": 1, "keywords" : 1}`.
  This query is valid for queries structured in the following order:

  – timestamp, retweet_count, keywords
  – timestamp
  – timestamp, retweet_count

  This query is **not** valid for queries structured in the following order:

  – retweet_count, timestamp, keywords
  – keywords
  – timestamp, keywords

- **Indexes can contain, at most, one array.** Twitter provides Tweet metadata in the form of arrays, but we can only use one in any given index.

3.8 Extracting Documents: Retrieving All Documents in a Collection

The simplest query we can provide to MongoDB is to return all of the data in a collection. We use MongoDB's `find` function to do this, an example of which is shown in Listing 3.3.

3.9 Filtering Documents: Number of Tweets Generated in a Certain Hour

Suppose we want to know the number of Tweets in our dataset from a particular hour. To do this we will have to filter our data by the `timestamp` field with "operators": special values that act as functions in retrieving data.

Listing 3.4 shows how we can drill down to extract data only from this hour. We use the `$gt` ("greater than"), and `$lte` ("less than or equal to") operators to pull dates from this time range. Notice that there is no explicit "AND" or "OR" operator specified. MongoDB treats all co-occurring key/value pairs as "AND"s unless explicitly specified by the `$or` operator.[5] Finally, the result of this query is passed to the `count` function, which returns the number of documents returned by the `find` function.

---

[5]For more operators, see http://docs.mongodb.org/manual/reference/operator/.

Listing 3.3 Get all of the Tweets in a collection

```
> db.tweets.find()
{ "_id" : ObjectId("51e6d70cd13954bd0dd9e09d"), ... }
{ "_id" : ObjectId("51e6d70cd13954bd0dd9e09e"), ... }
...
has more
```
Source: Chapter3/find_all_tweets.js

Listing 3.4 Get all of the Tweets from a single hour

```
> var NOVEMBER = 10; //Months are zero-indexed.
> var query = {
...       "timestamp" : {
...           "$gte": +new Date(2011, NOVEMBER, 15, 10),
...           "$lt": +new Date(2011, NOVEMBER, 16, 11)
...       }
... };
> db.tweets.find(query).count();
22169
```
Source: Chapter3/tweets_from_one_hour.js

Listing 3.5 Sort Tweets by time published

```
> db.tweets.find().sort({"timestamp": -1})
{ "_id" : ObjectId("51e6d713d13954bd0ddaa097"), ... }
{ "_id" : ObjectId("51e6d713d13954bd0ddaa096"), ... }
has more
```
Source: Chapter3/most_recent_tweets.js

3.10 Sorting Documents: Finding the Most Recent Tweets

To find the most recent Tweets, we will have to sort the data. MongoDB provides a sort function that will order the Tweet by a specified field. Listing 3.5 shows an example of how to use sort to order data by timestamp. Because we used "−1" in the value of the key value pair, MongoDB will return the data in descending order. For ascending order, use "1".

Without the index created in Sect. 3.7, we would have caused the error shown in Listing 3.6. Even with a relatively small collection, MongoDB cannot sort the data in a manageable amount of time, however with an index it is very fast.

Listing 3.6 Error generated without an index on "timestamp"

```
> db.tweets.find().sort({"timestamp": -1})
error: {
  "$err" : "too much data for sort() with no index.  add an
      index or specify a smaller limit",
  "code" : 10128
}
```

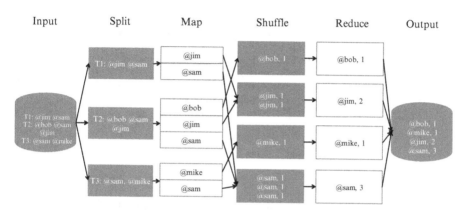

Fig. 3.3 MapReduce framework. The steps in white are implemented by the reader. MongoDB takes the documents from the database and runs each one through the map function. It then sorts the emitted keys and runs each key and its values through the reduce function. The output from the reduce function is stored in another collection

3.11 Grouping Documents: Identifying the Most Mentioned Users

With some simple use of the find and count functions, you can learn volumes about the data you have collected. However, when it comes to aggregating data, we will need to employ another set of functions, collectively called MapReduce.

MapReduce consists of two steps: "Map", and "Reduce". In the map step, data is extracted, filtered, and processed to be sent to the reduce function. The mapper processes in the map step emit a series of key/value pairs. These key value pairs are sorted, and the values associated with each unique key are sent to a reduce process. Each reduce process then computes a value for the key it is handed. A diagram for this process is shown in Fig. 3.3. The MongoDB code for this example is shown in Listing 3.7.

Listing 3.7 MapReduce function that lists the most mentioned users

```
> /*
> * This function extracts each user mentioned,
> * and the count of each mention.
> * The function takes 0 parameters, as the document
> * will be passed through context (the 'this' object).
> */
> var mapFunction = function(){
...        //loop through all of the mentions in the document.
...        var userMentions = this.entities.user_mentions;
...        for(var i = 0; i < userMentions.length; i++){
...            //check that the username is not blank.
...            if(userMentions[i].screen_name.length > 0){
...                //emit the username (key) and
...                //the count (value, in this case always 1).
...                emit(userMentions[i].screen_name, 1);
...            }
...        }
... }

> /*
> * This function sums the number of mentions of each user
> */
> var reduceFunction = function(keyUsername, occurs){
...        return Array.sum(occurs);
... }

> // Perform the MapReduce operation, and store the results
> // in a new collection, "most_mentioned_users".
> db.tweets.mapReduce(mapFunction, reduceFunction, {"out": "
    most_mentioned_users"});

> // List the top 5 most-mentioned users
> db.most_mentioned_users.find().sort({"value": -1}).limit(5)
{ "_id" : "MikeBloomberg", "value" : 727 }
{ "_id" : "OccupyWallSt", "value" : 588 }
{ "_id" : "OccupyWallStNYC", "value" : 428 }
{ "_id" : "JoshHarkinson", "value" : 295 }
{ "_id" : "ydanis", "value" : 260 }
Source: Chapter3/mapreduce.js
```

In Listing 3.7, the MapReduce is constructed as follows. The map function, called mapFunction, looks at each individual Tweet and pulls out the mentioned users. It then constructs the key/value pair to be sent to the reducer. The key is the user that was mentioned, and the value is 1. MongoDB then creates a unique reducer for each unique key and calls the reduce function, reduceFunction, on each key. The reducer then takes this list of values and calculates the sum. The result is a list of mentioned users and the count of the number of mentions for that user.

3.12   Further Reading

More information on MongoDB can be found in [2] and the MongoDB specification [1]. For more conversions between MongoDB's document-based syntax and SQL, see http://docs.mongodb.org/manual/reference/sql-comparison/. More information on other NoSQL implementations can be found in [3].

References

1. 10gen. The mongodb 2.4 manual. http://docs.mongodb.org/manual/, 2013.
2. K. Chodorow. *MongoDB: the definitive guide*. O'Reilly, 2013.
3. E. Redmond and J. R. Wilson. *Seven Databases in Seven Weeks*. Pragmatic Programmers, 2012.

Chapter 4
Analyzing Twitter Data

So far we have discussed the collection and management of a large set of Tweets. It is time to put these Tweets to work to gain information about the data we have collected. This chapter focuses on two key aspects of Twitter data for data analysis: networks and text.

When analyzing Twitter data, we can ask many questions. Who is the most important? What are people talking about? How are they responding to a product? In this chapter we will discuss how to answer these questions via data analysis.

4.1 Network Measures

Many of the questions that we ask of our Twitter data can be answered through network analysis. Questions such as "who is important?", "who talks to whom?", and "what is important?" can all be answered through a network. Using proper network measures, we can find these important actors or topics in a network.

4.1.1 What Is a Network?

A network[1] is a set of vertices linked by a set of edges. While this representation is very simple, the choices made when creating the network can make a huge difference in the way it is interpreted.

4.1.1.1 Vertices

Vertices are the elements that comprise a network. In both networks presented in Fig. 4.1, "Alice", "Bob", and "Carol" are the vertices in the network. Vertices can

---

[1] This is also commonly referred to as a *graph*.

S. Kumar et al., *Twitter Data Analytics*, SpringerBriefs in Computer Science, DOI 10.1007/978-1-4614-9372-3_4, © The Author(s) 2014

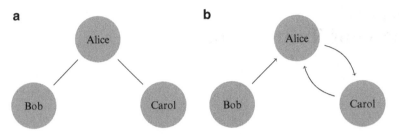

Fig. 4.1 Some basic networks illustrating the different kind of edges in a network. (**a**) A simple undirected network. Here, links are symmetric or nondirectional (Alice is joined to Bob and vice versa). (**b**) A simple directed network. Here, links are asymmetric or directional (Bob is joined to Alice but Alice is *not* joined to Bob)

represent literally *anything*. Starting with the definition of vertices as "users", we will present some basic concepts. In future sections we will show how different definitions for vertices can lead to different questions from the data.

4.1.1.2   Edges

Simply put, edges join vertices. There are several different types of edges that can join vertices. Undirected edges join vertices in a symmetric fashion. If two people engage in a conversation, then this is best represented by an undirected edge because for User A to converse with User B, User B must also converse with User A.

Contrary to undirected edges are directed edges, which signify a one-way relationship between two users. A retweet relationship can be thought of as a directed edge because User A can retweet User B without User B reciprocating.

Another property to consider is "edge weight". The weight of an edge signifies its importance when compared to other edges. One natural usage of weighted edges is in a user mention network, where the weight is the number of times the User A mentions User B.

4.1.1.3   Paths

A *path* is a sequence of nodes connected by a sequence of edges. A path always starts at one vertex, and ends at another vertex. We follow a path by starting at the first node in the sequence and traveling to each subsequent node in order. If no edge exists between two adjacent nodes in the sequence, then the path is invalid. In Fig. 4.1b, a path exists from Bob to Carol, however a path does not exist from Carol to Bob (directed edges cannot be followed in reverse order).

In addition to paths, we also have *shortest paths*. While there may be many ways to get from one node to another, the shortest path is the shortest sequence of nodes. It is important to find the shortest path from one node to another, as information will most often travel along this path.

4.1.2 Networks from Twitter Data

In the previous section we discussed the constructs that make up a network. We have kept our definition of a network general to allow for flexibility in the interpretation of the calculations performed on these networks when building them with different definitions in mind. When we build a network, our definition of nodes and edges determines the meaning of the measures performed. We begin by constructing a retweet network.

The retweet network is special because it can be derived directly through the Streaming API, there is no need to wait on the highly-limited REST APIs to extract this network. Additionally, it carries a very intuitive meaning. The nodes are individual users. The *directed* edges indicate the flow of information in the network. An edge from node A to node B indicates that A has retweeted B, forwarding B's information to his followers. Studying the structure of this network yields information about the ways the users communicate, and how highly they value each others' information.

4.1.3 Centrality: Who Is Important?

Often, we want to know who is the most important person in the network. The question of importance may not be as straightforward as it seems. There are several dimensions along which one may be considered important. Measures of importance in social networks are called "centrality measures". Here, we discuss the three that are used most frequently in social media analysis. Each provides a different view of who is important in the network. We will use the retweet network to demonstrate these centrality concepts.

4.1.3.1 Degree Centrality: Who Gets the Most Retweets?

One of the most commonly-used centrality measures is degree centrality. The calculation is simple: count the number of links attached to the node, this is their degree centrality. In the case of a directed graph, there are two interpretations: *In-Degree Centrality*, the number of edges entering the node, and *Out-Degree*, the number of edges leaving the node. In the context of our retweet networks, *In-Degree Centrality* is the number of users that retweet the node, and *Out-Degree Centrality* is the number of users the node retweets.[2] In Listing 4.1, we present an example of *In-Degree Centrality*. In Fig. 4.2, we see an example retweet graph with the nodes sized by their *In-Degree Centrality*. Interpreting this picture is simple, since Alice has the most edges pointing towards her, she is the most important node in the network.

---

[2]Notice that we have omitted an edge weight based upon number of times a retweet occurs. We omit this for simplicity.

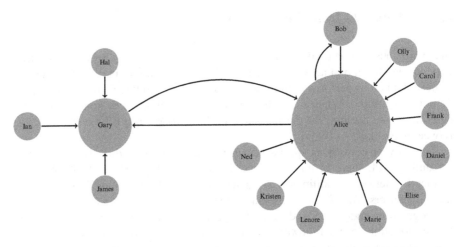

Fig. 4.2 In-Degree Centrality. Alice, who has the most users retweeting her, has the highest In-Degree Centrality

---

Listing 4.1 In-Degree Centrality calculation

```
...
//The graph representation in JUNG.
private Hypergraph<T, ?> graph;
...
/**
 * @return The In-Degree Centrality of the vertex.
 */
public Double getVertexScore(T node) {
  return (double) graph.getInEdges(node).size();
}
}
```
Source: Chapter4/util/InDegreeScorer.java

Listing 4.2 Eigenvector Centrality calculation

```
...
public EigenVectorScorer(Hypergraph<UserNode,
    RetweetEdge> graph){
  users = new UserNode[graph.getVertexCount()];
  graph.getVertices().toArray(users);

  /* Step 1: Create the adjacency matrix.
   *
   * An adjacency matrix is a matrix with N users and N
   *    columns,
   * where N is the number of nodes in the network.
   * An entry in the matrix is 1 when node i is joined to node
   *    j,
   * and 0 otherwise.
   */
```

```
SparseDoubleMatrix2D matrix =
  new SparseDoubleMatrix2D(users.length, users.length);
for(int i = 0; i < users.length; i++){
  for(int j = 0; j < users.length; j++){
    matrix.setQuick(i, j, graph.containsEdge(new RetweetEdge
      (users[i], users[j])) ? 1 : 0);
  }
}


/* Step 2: Find the principle eigenvector.
 * For more information on eigen-decomposition please see
 * \url{http://mathworld.wolfram.com/EigenDecomposition.html}
 */
EigenvalueDecomposition eig = new EigenvalueDecomposition(
    matrix);
DoubleMatrix2D eigenVals = eig.getD();
eigenVectors = eig.getV();

dominantEigenvectorIdx = 0;
for(int i = 1; i < eigenVals.columns(); i++){
  if(eigenVals.getQuick(dominantEigenvectorIdx,
    dominantEigenvectorIdx) <
    eigenVals.getQuick(i, i)){
    dominantEigenvectorIdx = i;
  }
 }
 }
 ...
}
```
Source: Chapter4/util/EigenVectorScorer.java

4.1.3.2   Eigenvector Centrality: Who Is the Most Influential?

With Degree Centrality the key question was "how many people retweeted this node?" *Eigenvector Centrality* builds upon this to ask "how important are these retweeters?" Figure 4.3 shows the same network as before, this time with nodes scaled by Eigenvector Centrality. We see that Bob, largely ignored by Degree Centrality, is the most important node through the lens of Eigenvector Centrality. This is because Alice, a high-degree node, gets information from Bob. An example of this calculation is shown in Listing 4.2.

4.1.3.3   Betweenness Centrality: Who Controls the Flow of Information?

Here we view importance from another perspective: the user's ability to control the flow of information. When information travels through a network, it takes the most convenient path possible. The most convenient path in a network is the *shortest path*. Betweenness centrality measures the number of shortest paths in which the user is in the sequence of nodes in the path.

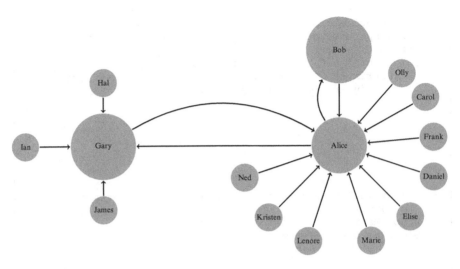

Fig. 4.3 Eigenvector Centrality. Alice, the user with the most retweets, listens to Bob, elevating his centrality

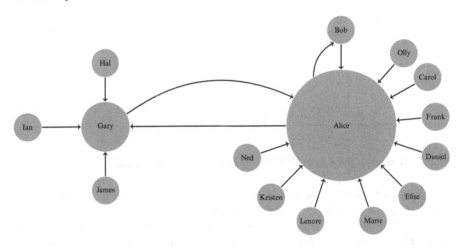

Fig. 4.4 Betweenness Centrality. Alice is the most important because most of the shortest paths go through her

In this measure of centrality, a user is important because he controls the routes of information flow in the network. His centrality score is the fraction of shortest paths that travel through the node. An example of the calculation is shown in Listing 4.3. Figure 4.4 shows an example retweet graph with the nodes sized by their *Betweenness Centrality*. This time, Alice is not important because she has the most adjacent nodes, but because many of the shortest paths go through her.

Listing 4.3 Betweenness Centrality calculation

```
public class BetweennessCentralityExample {
  public static void main(String[] args){

    File tweetFile;

    if(args.length > 0){
      tweetFile = new File(args[0]);
    }
    else{
      tweetFile = new File("synthetic_retweet_network.json");
    }

    DirectedGraph<UserNode, RetweetEdge> retweetGraph =
        TweetFileToGraph.getRetweetNetwork(tweetFile);

    //calculate the betweenness centrality
    BetweennessCentrality<UserNode, RetweetEdge> betweenness =
        new BetweennessCentrality<UserNode, RetweetEdge>(
        retweetGraph);

    betweenness.evaluate();
    betweenness.printRankings(true, true);

  }
}
```
Source: Chapter4/util/BetweennessScorer.java

4.1.4   Finding Related Information with Networks

All of our network constructions so far have only considered users as nodes and
edges as retweets. We can choose any object as a node and any relation as an edge.
Let's take a look at another network construction that allows us to ask different
questions about our Twitter data.

What if we wanted to see how hashtags are related? There are many valid ways
to measure this, but, true to this chapter, we will measure this using a network-based
approach. We will consider a new network construction where nodes are individual
hashtags and edges are hashtags that co-occur within the same Tweet. We will
weight the edges by the number of times the hashtags co-occur in a Tweet. Because
we do not care about the hashtag order, the edges are not directed. Figure 4.5 shows
an example hashtag network that can be constructed from the topics discussed in
our Occupy Wall Street dataset.

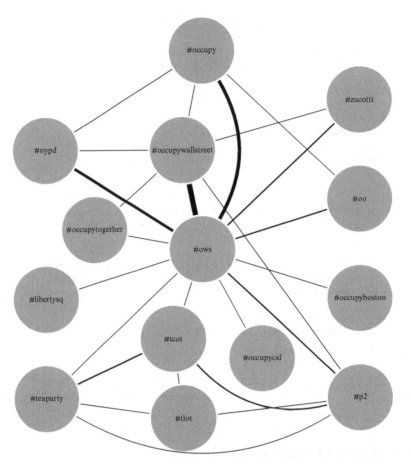

Fig. 4.5 Network of co-occurrence of hashtags within Tweets. Edge weight represents number of co-occurrences

4.2 Text Measures

Previously, we explored a model that exploits the links between the entities to help us find the key players in the data. Here, we will focus on the Tweet's text to better understand what the users are talking about. We move away from the network model we've used previously and discuss other methods for text analysis. We first explore topic modeling, an approach that finds natural topics within the text. We then move on to sentiment analysis, the practice of associating a document with a sentiment score.

Table 4.1 An example of the most significant words from one topic from LDA

Word	Probability (%)
ows	8.0
nypd	2.0
occupywallstreet	1.0
park	1.0
occupi	1.0
protest	1.0
nyc	1.0
evict	1.0
citi	1.0
polic	1.0
zuccotti	1.0
...	...

4.2.1 Finding Topics in the Text

The data we collect from Twitter quickly grows to immense proportions. In fact, they grow so large that attempting to read each individual Tweet quickly becomes a hopeless cause. A more reachable goal is to get a high-level understanding of what our users are talking about. One way to do this is by understanding the *topics* the users are discussing in their Tweets. In this section we discuss the automatic discovery of topics in the text through "topic modeling" with latent Dirichlet allocation (LDA), a popular topic modeling algorithm.

4.2.1.1 What Is a Topic?

Every topic in LDA is a collection of words. Each topic contains all of the words in the corpus with a probability of the word belonging to that topic. So, while all of the words in the topic are the same, the weight they are given differs between topics. For example, we may find a topic related to sports that is made up of 40% "basketball", 35% "football", 15% "baseball", ..., 0.02% "congress", and 0.01% "Obama". Another topic related to politics could be made up of 35% "congress", 30% "Obama", ..., 1% "football", 0.1% "baseball", 0.1% "basketball". Because each topic contains every word we will only view the top words when inspecting a topic.

LDA finds the most probable words for a topic, associating each topic with a theme is left to the user. An example topic from the Occupy Wall Street data is shown in Table 4.1.

4.2.1.2 LDA Calculation with MALLET

To perform the LDA computation in Java, we use the MALLET©[3] library. Listing 4.4 shows the computation in MALLET. As we can see, most of the work is done for us, the real effort is in the preprocessing of the documents. To get the documents ready for LDA, we define a preprocessing pipeline that processes each document. We'll enumerate our preprocessing pipeline:

1. **Lowercase** – Strip casing off of all words in the document. "No more media blackout hiding #OCCUPYWALLSTREET! :)" becomes "no more media blackout hiding #occupywallstreet! :)".
2. **Tokenize** – Convert the string to a list of tokens based on whitespace. This process also removes punctuation marks from the text. This becomes the list [no, more, media, blackout, hiding, #occupywallstreet].
3. **Stopword Removal** – Remove "stopwords", words so common that their presence does not tell us anything about the dataset. [no, media, blackout, hiding, #occupywallstreet].
4. **Stemming** – Reduce each word to its stem, removing any prefixes or suffixes. [no, media, blackout, hide, #occupywallstreet].
5. **Vectorization** – Convert the sequence of words to a vector that, instead of containing the words, contains a sequence of numbers for each word in the vocabulary. The value at each index corresponds to the number of times each word appears in the document.

Listing 4.4 LDA computation with MALLET

```
...
private static final String STOP_WORDS = "stopwords.txt";
private static final int ITERATIONS = 100;
private static final int THREADS = 4;
private static final int NUM_TOPICS = 25;
private static final int NUM_WORDS_TO_ANALYZE = 25;


...
// Lowercase, tokenize, remove stopwords, and convert to
    features
pipeList.add((Pipe) new CharSequenceLowercase());
pipeList.add((Pipe) new CharSequence2TokenSequence(Pattern.
    compile("\\p{L}[\\p{L}\\p{P}]+\\p{L}")));
pipeList.add((Pipe) new TokenSequenceRemoveStopwords(
    stopwords, "UTF-8", false, false, false));
pipeList.add((Pipe) new PorterStemmer());
pipeList.add((Pipe) new TokenSequence2FeatureSequence());

InstanceList instances = new InstanceList(new SerialPipes(
    pipeList));
```

---

[3]http://mallet.cs.umass.edu/

```
...
instances.addThruPipe(new StringArrayIterator(textList.
    toArray(new String[textList.size()]))));

ParallelTopicModel model = new ParallelTopicModel
    (NUM_TOPICS);
model.addInstances(instances);
model.setNumThreads(THREADS);
model.setNumIterations(ITERATIONS);
model.estimate();

// The data alphabet maps word IDs to strings
Alphabet dataAlphabet = instances.getDataAlphabet();

int topicIdx=0;
StringBuilder sb;
for (TreeSet<IDSorter> set : model.getSortedWords()) {
    sb = new StringBuilder().append(topicIdx);
    sb.append(" - ");
    int j = 0;
    double sum = 0.0;
    for (IDSorter s : set) {
      sum += s.getWeight();
    }
    for (IDSorter s : set) {
        sb.append(dataAlphabet.lookupObject(s.getID())).
            append(":").append(s.getWeight() / sum).
            append(", ");
        if (++j >= NUM_WORDS_TO_ANALYZE) break;
    }
    System.out.println(sb.append("\n").toString());
    topicIdx++;
  }
 }
}
```
Source: Chapter4/tweetlda/LDA.java

4.2.2 Sentiment Analysis

Often its not important to know what users are saying, but *how they are saying it*. "Sentiment analysis" seeks to automatically associate a piece of text with a "sentiment score", a positive or negative emotional score. Aggregating sentiment can give an idea of how people are responding to a company, product, or topic.

4.2.2.1 Sentiment Analysis Overview

Sentiment analysis, is done on a per-Tweet basis. The words in each Tweet are compared with those in other Tweets that have been previously labeled as "positive",

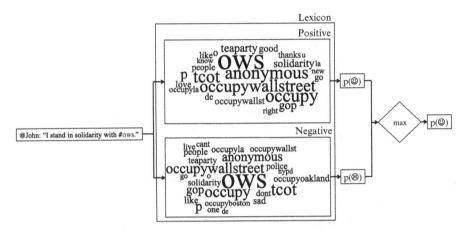

Fig. 4.6 The sentiment analysis workflow. John's Tweet is compared against a lexicon of words and their likelihood to be positive/negative. The most probable label is then taken as that Tweet's sentiment

or "negative". After looking at these words, the algorithm then judges whether the text in the Tweet is positive or negative based on the likelihood for each possibility. A workflow is shown in Fig. 4.6.

To compare the content in the Tweets, we must first find a *lexicon*, a dictionary of words and their positive and negative scores.[4] When choosing a sentiment lexicon, we need to be careful about the source used to build it. Words have different sentiments in different contexts. For example, in a lexicon built by looking at movie reviews, "bomb" would likely have a positive sentiment ("that movie was the bomb"). In a lexicon built by looking at world news articles, "bomb" would likely be negative ("the bomb detonated in...").

The sentiment analysis algorithm we use in this book is based on a Naïve Bayes Classifier. It classifies a Tweet as positive or negative by comparing each word in the Tweet with the labeled words in the lexicon. If the words in the Tweet have been used more in positive Tweets, then the Tweet is labeled as positive. On the other hand, if the words in the Tweet have been associated more with negative Tweets, then the Tweet is labeled as negative.

4.2.2.2 Building a Lexicon Automatically

To get around the potential issue of having an unsuitable lexicon, we will construct our lexicon automatically for each dataset. Because we are using data collected directly from Twitter, we do not have explicit "positive", or "negative" labels.

---

[4]Some sentiment lexicons are available for free, such as SentiWordNet (http://sentiwordnet.isti.cnr.it/).

Instead, we use Tweets that contain emoticons as labeled data. We will mark Tweets that contain ":)", ":D", or similar as positive, and Tweets that contain ":(", ";-(", or similar as negative. In this way we know that the lexicon is built using relevant data.

Figure 4.6 shows the top 25 most likely words for each sentiment in the Occupy Wall Street dataset. This gives us an idea of what words define each sentiment. When reading these word clouds, one might get confused about the prominence of "ows" in both groups. While it is the most prominent word for both sentiments, its appearance in the Tweet does nothing to help us understand its sentiment. It is more important to look for words that appear in one sentiment class but *not the other*, or those with a large size difference.

4.2.2.3   The Sentiment Analysis Process

We have outlined the process to create a sentiment analysis framework. Listing 4.5 contains a snippet that performs the sentiment analysis task. The code begins by enumerating each Tweet in the dataset, building a lexicon from the Tweets that use an emoticon. Next, it enumerates the Tweets again, calculating a sentiment score for each Tweet that does *not* have an emoticon. Listing 4.5 shows an example of this process. For the code that actually builds the lexicon and calculates the sentiment score, see `NaiveBayesSentimentClassifier.java`.

Listing 4.5 Sentiment analysis runner

```
public class TestNBC {
  public static void main(String[] args){

    String filename = args.length >= 1 ? args[0] :
      "testows.json";

    //initialize the sentiment classifier
    NaiveBayesSentimentClassifier nbsc = new
      NaiveBayesSentimentClassifier();

    try {
      //read the file, and train each document
      JsonStreamParser parser =
        new JsonStreamParser(new FileReader(filename));
      JsonObject elem;
      String text;
      while (parser.hasNext()) {
          elem = parser.next().getAsJsonObject();
          text = elem.get("text").getAsString();
          nbsc.trainInstance(text);
      }

        //now go through and classify each line as positive or
            negative
      parser =
```

```
            new JsonStreamParser(new FileReader(filename));
        while (parser.hasNext()) {
                elem = parser.next().getAsJsonObject();
                text = elem.get("text").getAsString();
                Classification c = nbsc.classify(text);
                System.out.println(c + " -> " + text);
        }

        ...

    }
}
```
Source: Chapter4/classification/bayes/TestNBC.java

4.3 Further Reading

For a detailed introduction to the field of network analysis, we refer the reader to [3,5,7].

Another introduction to LDA by Edwin Chen, a former engineer at Twitter, can be found on his blog [2]. For a deeper review of LDA, the reader can consult [1,4].

Many approaches have been taken towards sentiment analysis. For an overview of the field, see [6].

References

1. D. M. Blei, A. Y. Ng, and M. I. Jordan. Latent dirichlet allocation. *The Journal of Machine Learning Research*, 3:993–1022, 2003.
2. E. Chen. Introduction to Latent Dirichlet Allocation. http://blog.echen.me/2011/08/22/introduction-to-latent-dirichlet-allocation/, August 2011.
3. M. Hennig, U. Brandes, J. Pfeffer, and I. Mergel. *Studying Social Networks: A Guide to Empirical Research*. Campus Verlag, 2012.
4. K. Murphy. *Machine Learning: A Probabilistic Perspective*. Adaptive computation and machine learning series. MIT Press, 2012.
5. M. Newman. *Networks: An Introduction*. OUP Oxford, 2009.
6. B. Pang and L. Lee. Opinion Mining and Sentiment Analysis. *Found. Trends Inf. Retr.*, 2(1–2):1–135, Jan. 2008.
7. R. Zafarani, M.-A. Abbasi, and H. Liu. *Social Media Mining: An Introduction*. Cambridge University Press, Forthcoming.

Chapter 5
Visualizing Twitter Data

When users interact on Twitter, network information is generated, and when they publish Tweets, textual information is generated. Tweets themselves have other embedded information, such as location information. In addition, users have profiles where they describe themselves through fields, such as their name and website. Visualization techniques can help us efficiently analyze and understand how and why users interact on Twitter. In this chapter, we discuss techniques to create visualizations for the four types of information: network, temporal, geospatial, and textual information. While discussing the techniques, we follow the visualization mantra: "Overview first, then zoom and filter. Details on demand" [4].

5.1 Visualizing Network Information

In the previous chapter, we discussed network measures to identify important people and concepts in the network. In this section, we will continue that discussion and present a technique to visualize a network to gain insight into how and why users interact. We will focus our discussion on two types of networks:

- Information flow networks, and
- Friend-Follower networks.

Below, we discuss each type of networks in detail and provide an example to illustrate the unique aspects of the network and how visualization can help in understanding them.

5.1.1 Information Flow Networks

On Twitter, information spreads primarily through retweeting. The resulting Tweet is called a retweet. When we visualize retweets we are essentially visualizing the flow of information in the network. In the previous chapter, we discussed centrality

S. Kumar et al., *Twitter Data Analytics*, SpringerBriefs in Computer Science,
DOI 10.1007/978-1-4614-9372-3__5, © The Author(s) 2014

measures used to identify important nodes. By visualizing the network, we can aid in further analysis of important nodes, by identifying information propagation paths, as well as observing the interaction between information producers and consumers, which cannot be effectively conveyed by measures alone.

Retweets are marked by the characteristic prefix "RT" followed by the name of the user who originally published the Tweet. For example, consider the following Tweet published by the user John:

> RT @Peter: Full-time: Chelsea 3-1 Steaua Bucharest. (3-2 on agg) and we're through to the quarter-finals. #CFC

Here "Peter" is the name of the user who originally posted the Tweet, which was retweeted by "John". Using this information, an information flow network can be created by connecting information producers with information consumers. When collecting Tweets, one can identify retweets by checking for the presence of the element "retweeted_status" in the JSON response. As an example, a part of the element is presented in Listing 5.1.

5.1.1.1 Retweet Network Fallacy

An important yet subtle property of this network is that one can only identify the original source of the information and not the intermediate users along the information propagation path. For example consider the above example and imagine two users "Alice" and "Bob" who retweet the Tweet from "John". The "retweeted_status" element of this Tweet will contain the original producer of the Tweet i.e., "Peter" as the source of the Tweet. In Fig. 5.1a. We can see that the tweet from "Peter" is retweeted by "John", which is subsequently retweeted by "Alice" and "Bob". However, when these Tweets are collected via the Twitter APIs, we only observe the propagation path seen in Fig. 5.1b. This means that we cannot identify the full path of propagation, but only the source and the destination.

Listing 5.1 Retweet object inside a Tweet object

```
{
    //Other Tweet elements
    . . .
    "created_at": "Thu Mar 14 23:25:03 +0000 2013",
    "text": "RT @Peter: Full-time: Chelsea 3-1 Steaua Bucharest
        . (3-2 on agg) and we're through to the quarter-finals.
        #CFC",
    "retweeted_status": {
        "text": "Full-time: Chelsea 3-1 Steaua Bucharest. (3-2
            on agg) and we're through to the quarter-finals. #
            CFC",
        "retweeted": true,
        . . .
        //other retweet elements
}
```

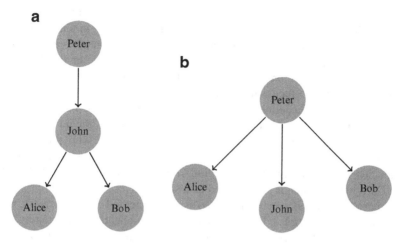

Fig. 5.1 Perceived and actual information propagation path of Tweets on Twitter. (**a**) Actual propagation path. (**b**) Path extracted from Twitter API

5.1.1.2   Visualizing Retweet Propagation

A popular technique to visualize graphs is the *force-directed layout.* We will now briefly explain how this layout can be used to visualize information propagation networks.

Force-Directed Layout: To clearly identify the information propagation in a graph, the nodes must be spread out, especially when the number of nodes in the graph is large. The force-directed layout tackles the problem of placing nodes within a restricted space while simultaneously ensuring that the resulting layout is aesthetically pleasing. Force-directed layout is intuitive and suitable for network graphs extracted from Twitter. There are two principles in graph drawing:

- Nodes connected to each other should be placed as close to each other as possible.
- Nodes should not be drawn too close to each other.

In the force-directed layout, every node exerts a repelling force on every other node. The edges act as springs and exert an attractive force between connected nodes. Typically, the layout starts with a random placement of the nodes in the space. The constraint of keeping connected nodes close to each other ensures that the layout is also compact. In the example discussed here, we will use the implementation in D3.[1]

Applying the Force-Directed Layout to Retweet Network: Visualizing large networks generated from the Tweets is quite expensive and can lead to "spaghetti" networks, which are not very useful as they can consist of too many nodes and edges. For a single Tweet alone, the retweet network could consist of thousands of users. The visualization can be simplified and simultaneously made more

---

[1]https://github.com/mbostock/d3/wiki/Force-Layout

interesting by adding context to the network. During an event, a user would want to analyze the event from different perspectives. For example, in a natural disaster like a thunderstorm, instead of analyzing all the retweets related to the disaster, a first responder might be more interested in reports of damage or flooding. Thus, by filtering the information, we can make the visualization manageable while enhancing the ability to focus on topics of interest.

To achieve this, we define groups of words called topics. For example, we create topic 1 with "#zuccotti" (Light) and topic 2 with "#nypd" (Dark). We can now generate a retweet network consisting of people who retweeted text matching these topics. Before we visualize the network, we must first extract and format it. This can be done using the method *ConvertTweetsToDiffusionPath* in class *CreateD3Network*, which is summarized in Listing 5.2.

The extracted network can be visualized using the method *create_network*, which is summarized in Listing 5.3. Figure 5.2 shows the visualization of the top five most frequently retweeted nodes and those who retweeted them on topics 1 and 2. The size of a node indicates its importance in the network. Larger nodes have been retweeted more often than smaller nodes. Nodes are colored according to their topic preference. The links are directed and show the flow of information. Here, not only can we identify important information producers (large nodes) as well as information consumers (nodes with a large number of inlinks). Additionally, the network shows that people retweet across topics, which is evident from the connections between the users.

Listing 5.2 Extracting the retweet network

```
public JSONObject ConvertTweetsToDiffusionPath(String inFilename
    ,int numNodeClasses, JSONObject hashtags, int num_nodes) {
        //Step 1: Read through the file and process Tweets
            matching the topics
            . . .
        //Step 2: Identify the size of the nodes based on the
            number of times they are retweeted
        ArrayList<NetworkNode> nodes = ComputeGroupsSqrt(
            returnnodes, max, min, numNodeClasses);
            . . .
        /** Step 3
            * Prune the network to keep only the top |
                nodes_to_visit| nodes in the network.
            * Recursively visit all top nodes and retain their
                connections.
            */
        for(int k=0;k<nodes_to_visit;k++) {
            NetworkNode nd = nodes.get(k);
            nd.level = 0;
            HashMap<String,NetworkNode> rtnodes =
                GetNextHopConnections(userconnections,nd,new
                HashMap<String,NetworkNode>());
            . . .
        /** Step 4: Compact the nodes of the network by removing
            * all nodes who have never been retweeted
            */
        Set<String> allnodes = prunednodes.keySet();
```

```
    //Store the list of retweeted nodes
    ArrayList<NetworkNode> finalnodes = new ArrayList<
        NetworkNode>();
    for(String n:allnodes) {
        . . .
    }
    //Sort in ascending order of the Node ID
    Collections.sort(finalnodes,new NodeIDComparator());
    //Step 5: Reformat the network into  D3 format
    return GetD3Structure(finalnodes);
}
```
Source: Chapter5/network/CreateD3Network.java

Listing 5.3 Visualizing the retweet network using force-directed layout

```
create_network: function() {
  . . .
// Step 1: Create the nodes and the links in the network
network_page.net = network_page.create_dataset(network_page.
    jsondata, network_page.net, network_page.getGroup);
/**
  * Step 2: Initialize the D3 layout with the data and node
      settings that
  * define the forces acting on the nodes and start the layout
  */
network_page.force = d3.layout.force()
      .nodes(network_page.net.nodes)
      .links(network_page.net.links)
      .size([width,height])
      .charge(-500)
      .linkDistance(80)
      .theta(0.8)
      .gravity(0.2)
      .start();
  . . .
/**
  * Step 3: Compute the distance between nodes after each
      iteration
  * the forces are computed using the tick event.
  */
  network_page.force.on("tick", function() {
    link.attr("x1", function(d) { return d.source.x; })
        .attr("y1", function(d) { return d.source.y; })
        .attr("x2", function(d) { return d.target.x; })
        .attr("y2", function(d) { return d.target.y; });
    node.attr("cx", function(d) {
          return d.x = Math.max(r, Math.min(width - r, d.x));
        }).attr("cy", function(d) {
          return d.y = Math.max(r, Math.min(height - r, d.y));
        });
  }
}
```
Source: TwitterDataAnalytics/js/network.js

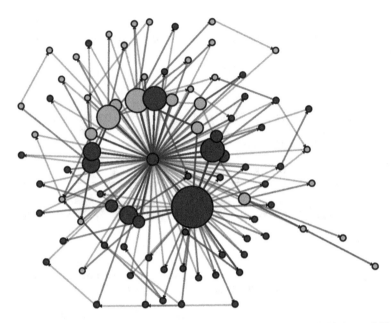

Fig. 5.2 A retweet network containing nodes from the topics "#zuccotti" (Light) and "#nypd" (Dark)

Adding context to the network: The network in Fig. 5.2 describes the relationship between nodes, but does not provide sufficient. Here, Tweets can help in further investigation of the role of the user in the event. They can also be used to identify parts of the network, which were instrumental in propagating relevant information. This is relevant in applications such as marketing, where a customer would be interested in maximizing the spread of a message at minimum cost.

Figure 5.3 shows the information panel created when the largest node in Fig. 5.2 is clicked. The panel shows the users who retweeted his Tweet. Each unique retweet is identified by a color and the aggregation of the number of retweets for each unique Tweet is presented in the form of a pie chart on the bottom of the panel highlighting the most retweeted Tweets.

5.1.2   Friend-Follower Networks

A friend-follower network consists of users as nodes and the edges describe *who follows whom*. Each user on Twitter is able to create two types of explicit relationships with another Twitter user. Suppose we have two users: Alice and Bob. If Alice starts following Bob, then we say that Alice is a "follower" of Bob. Here, the decision to initiate the connection is made by Alice. Bob is known as the "friend" of Alice.

The friends and followers of a user can be extracted directly using the Twitter REST APIs discussed in Chap. 2. Applying the force-directed layout would separate the connected component from the rest of the group.

Fig. 5.3 Information panel
showing the Tweets from
most retweeted user

5960662372693928329

- @3833960250514607362: call #nypd at 646-610-5000 and respectfully demand 1st amendment rights!! #ows #occuppywallstreet
- @1297829963558188053: update on #libertysq being raided by #nypd http//t.co/1qxc7gp6 #ows
- @1990034068600533934: call #nypd at 646-610-5000 and respectfully demand 1st amendment rights!! #ows #occuppywallstreet
- @1632978510408923869: city council member ydanis rodriguez beaten by #nypd and bleeding from head. #ows
- @1441465728317042272: city council member ydanis rodriguez beaten by #nypd and bleeding from head. #ows
- @2755634033224588482: city council member ydanis rodriguez beaten by #nypd and bleeding from head. #ows
- @2755634033224588482: city council member ydanis rodriguez beaten by #nypd and bleeding from head. #ows
- @4532294461120672775: city council member ydanis rodriguez beaten by #nypd and bleeding from head. #ows
- @4532294461120672775: city council member ydanis rodriguez beaten by #nypd and bleeding from head. #ows
- @4532294461120672775: sigh... city council member ydanis rodriguez beaten by #nypd and bleeding from head.
...
- @5306236007513202898: city council member ydanis rodriguez beaten by #nypd and bleeding from head. #ows
- @4532294461120672775: !? city council member ydanis rodriguez beaten by #nypd and bleeding from head. #ows
- @7403477541149523926: what? #nypd sawing down trees in #libertysq #ows / whiskey tango foxtrot. over.

5.2   Visualizing Temporal Information

Time-series data is time sensitive information about a variable. Such variables in
the case of Twitter include the volume of Tweets and daily interactions between
users. Time-series are also referred to as trends. Visualizing trends helps us detect
temporal patterns in the data, such as the periodic activity of users, which can help
us understand their actions on Twitter.

Time-series visualization can be used to

- Analyze information associated with time, and
- Present a natural ordering of time-oriented information.

Time-series visualization is typically a chart with one axis (generally the
x-axis) representing time and the y-axis representing a measurement along another
dimension, such as the volume of Tweets. Figure 5.4 presents the natural representation of the volume of Tweets collected per minute on Nov 15 in the
sample dataset. The information needed to generate this trend can be extracted

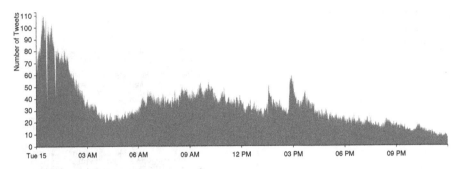

Fig. 5.4 Number of Tweets generated per minute on November 15, 2011

using the method *GenerateDataTrend* defined in the class *ExtractDatasetTrend* and summarized in Listing 5.4.

5.2.1   Extending the Capabilities of Trend Visualization

A simple trendline provides only an overview of the data and cannot support in-depth analysis. Operations such as zoom and filter can empower a user to drill-down and investigate the data more effectively. There are two techniques which can improve the user experience and add value to the visualization discussed above: brushing & linking and focus+context. Together, they can be used to facilitate drill-down analysis of information.

Brushing & Linking Brushing and linking is a technique to effectively use multiple views of the data to perform zoom and filter operations. In the collective action of brushing and linking, multiple views are linked to each other, with each view providing a different perspective of the data. Actions in the views are linked by means of user interaction. Selecting a region in one view simultaneously affects a change in the other view.

Focus+Context Focus+context is a technique to enable drilling down into the data. Trendlines provide an effective overview of some quantity, which in the case of Fig. 5.4, is the number of Tweets per minute. Focusing the view causes the view to zoom into the data to provide information in greater detail. Context is the action of de-magnifying to zoom out of the series at a particular point and it allows a user to obtain the overview of the data.

Listing 5.4 Extracting the time-series of the volume of Tweets

```java
// Time pattern used to count the volume of Tweets
final SimpleDateFormat SDM = new SimpleDateFormat("dd MMM yyyy
    HH:mm");

public JSONArray GenerateDataTrend(String inFilename) {
    HashMap<String,Integer> datecount = new HashMap<String,
        Integer>();
    // Step 1: Parse the time of publication of each Tweet and
        count the number of Tweets using SDM
        . . .
        /** DateInfo consists of a date string and the
            corresponding count.
         * It also implements a Comparator for sorting by time
         */
        ArrayList<DateInfo> dinfos = new ArrayList<DateInfo>();
        Set<String> keys = datecount.keySet();
        for(String key:keys) {
                DateInfo dinfo = new DateInfo();
                try {
                        dinfo.d = SDM.parse(key);
                } catch (ParseException ex) {
                        ex.printStackTrace();
                        continue;
                }
                dinfo.count = datecount.get(key);
                dinfos.add(dinfo);
    }
    // Step 2: Sort the counts in the increasing order of
        the time
    Collections.sort(dinfos);
    // Format and return the date string and the
        corresponding count
        . . .
}
```
Source: Chapter5/trends/ExtractDatasetTrend.java

An example of the two actions described above can be seen in Fig. 5.5. Initially the trend lines/views present the same overview of the trend as in Fig. 5.4. Brushing a region in the trendline at the bottom selects a time range. Since the two views are linked, this action causes the series on the top to focus into the selected time range, which in this case is 12–3 PM. The user can then observe the information in greater detail and identify finer patterns in the data. Removing the brushed region causes the original trend to re-appear and present the context. This helps a user analyze the overview of the time-series all the time and obtain details when necessary, which can be useful when investigating time-series data over extended periods of time. This visualization can be implemented using the function summarized in Listing 5.5, which is defined in *TrendLine.js*.

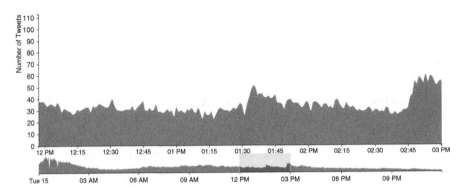

Fig. 5.5 A focus+context interface with brushing and linking

Listing 5.5 Extending trend lines with brushing and focus+context

```
window.onload = function() {
 // Step 1: Initialize the chart
   trendline.initialize();
 // Step 2: Fetch data using AJAX
           . . .
         //Step 3: Initialize the focus handlers
         trendline.focus.append("path").datum(data).attr("clip-
            path", "url(#clip)").attr("d", trendline.area);
         trendline.focus.append("g").attr("class", "x axis").
            attr("transform", "translate(0," + trendline.
            height + ")") .call(trendline.xAxis);
         trendline.focus.append("g").attr("class", "y axis").
            call(trendline.yAxis).append("text").attr("class",
             "ylabel").attr("text-anchor", "end").attr("y",
            -40).attr("dy", ".75em").attr("transform", "rotate
            (-90)").text("Number of Tweets");
         //Step 4: Context is initialized to series 2
         trendline.context.append("path").datum(data).attr("d",
            trendline.area2);
         trendline.context.append("g").attr("class", "x axis").
            attr("transform", "translate(0," + trendline.
            height2 + ")").call(trendline.xAxis2);
         // Step 5: Select the brushed region and focus on it
         trendline.context.append("g").attr("class",
         "x brush").call(trendline.brush).selectAll("rect").
            attr("y", -6).attr("height", trendline.height2 +
            7);
}
```
Source: TwitterDataAnalytics/js/trendLine.js

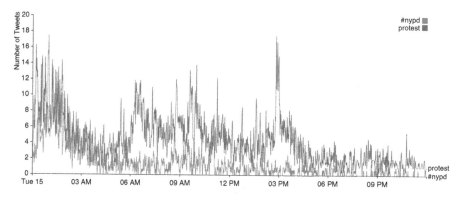

Fig. 5.6 Comparison of the volume of Tweets collected for "protest" and "#nypd"

5.2.2   Performing Comparisons of Time-Series Data

A typical application of trend lines is to perform comparative analysis. When dealing with time-oriented information, we often want to compare two or more quantities to gain a better understanding of the differences in temporal patterns. Time-series visualizations can help perform comparative analysis quickly with little effort from the user. As an example, we present a comparison of the word "protest" and the hashtag "#nypd" in Fig. 5.6.

The trendline shows that while both words were discussed during the initial hours of the day, the discussion of "#nypd" died quickly while the discussion of "protest" continued throughout the day. The chart can be generated using the method *GenerateGraph*, which is presented in Listing 5.6. The data for the chart can be extracted using the method *GenerateDataTrend* in class *TrendComparisonExample*. This method is similar to the one described in Listing 5.4, with the difference that each Tweet is first compared to the supplied words and only counted if it matches a word.

Listing 5.6 Generating trend lines to compare multiple time-series

```
function GenerateGraph(data) {
. . .
//Step 1: Initialize the chart
  trendcomp.svg.append("g")
          .attr("class", "x axis")
          .attr("transform", "translate(0," + trendcomp.height +
              ")")
          .call(trendcomp.xAxis);
  trendcomp.svg.append("g")
          .attr("class", "y axis")
          .call(trendcomp.yAxis)
          .append("text")
          .attr("transform", "rotate(-90)")
          .attr("class", "ylabel")
```

```
            .attr("y", 6)
            .attr("dy", ".71em")
            .style("text-anchor", "end")
            .text("# of Tweets");
// Step 2: Append the time series information for each word
  var word = trendcomp.svg.selectAll(".word")
            .data(words)
            .enter().append("g")
            .attr("class", "word");
  word.append("path")
            .attr("class", "line")
            .attr("d", function(d) { return trendcomp.line(d.
                values); })
            .style("stroke", function(d) { return trendcomp.color
                (d.word); });
  . . .
  //Step 3: Add the legend to the chart
  var legend = trendcomp.svg.selectAll(".legend")
            .data(trendcomp.color.domain().slice().reverse())
            .enter().append("g")
            .attr("class", "legend")
            .attr("transform", function(d, i) { return "translate
                (0," + i * 20 + ")"; });
  . . .
}
```
Source: TwitterDataAnalytics/js/trendComparison.js

Listing 5.7 Creating sparklines to compare multiple time-series

```
GenerateSparkLines:function(data) {
        //loop through the data and load sparkline for each word
        for(key in data) {
                $("#vizpanel").append(this.CreateTextElement(
                    key)).append(this.CreateSpanElement(key));
                $("span[data='"+key+"']").sparkline(data[key]);
        }
}
```
Source: TwitterDataAnalytics/js/sparkLine.js

5.2.2.1   Sparklines

Sparklines are simple and typically small graphics. They are a miniaturized version of the full trend line designed to provide a quick overview of the variation in a quantity over time. They are designed to be displayed along with the text that describes them, so they are minimalistic. For example, #nypd is a sparkline for #nypd. Sparklines can be embedded in text as well as other visualizations to summarize time-series information. Sparklines can be created using the method *GenerateSparkLines*, which is summarized in Listing 5.7.

Small multiples: This refers to the concept of creating a series of mini charts which can be used for quick comparison and summarization of information.

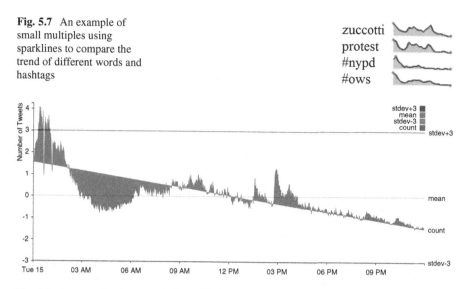

Fig. 5.7 An example of small multiples using sparklines to compare the trend of different words and hashtags

zuccotti

protest

#nypd

#ows

Fig. 5.8 An example of a control chart with the control line set to 3 standard deviations

Here, sparklines can be efficiently used to generate trending information of multiple entities for quick analysis, as in Fig. 5.7.

5.2.2.2   Control Charts

A control chart is a statistical tool used to detect abnormal variations in a process. This task is performed by measuring the stability of the process through the use of control limits. A control limit is a threshold which helps a user detect anomalous periods of activity. If the data falls within the control limits, then the process is considered stable. 3-standard deviations is typically chosen as the control limit. If the activity falls outside the 3-standard deviations, then it is considered abnormal and worthy of investigation.

As an example, let's look at Fig. 5.8. Here, we present the volume of Tweets generated every minute. Each value in the distribution is subtracted from the mean and the difference is divided by the standard deviation to center as well as scale the distribution using the methods in Listing 5.8. Therefore, the distribution has a mean 0 and a standard deviation of 1. The lower and upper control limit are set to 3 standard deviations. On Twitter, this could be used to detect events by

Listing 5.8 Methods to calculate mean and standard deviation

```
public double GetStandardDev(ArrayList<DateInfo> dateinfos,
     double mean) {
          double intsum = 0;
          int numperiods = dateinfos.size();
```

```
            for(DateInfo dinfo:dateinfos)
            {
                intsum+=Math.pow((dinfo.count - mean),2);
            }
            return Math.sqrt((double)intsum/numperiods);
}

public double GetMean(ArrayList<DateInfo> dateinfos) {
            int numperiods = dateinfos.size();
            int sum = 0;
            for(DateInfo dinfo:dateinfos)
            {
                sum +=dinfo.count;
            }
            return ((double)sum/numperiods);
}
```
Source: Chapter5/trends/ControlChartExample.java

monitoring the times at which the traffic exceeds the upper control limit or falls below the lower control limit. A user can then be notified, so the event can be further investigated. In Fig. 5.8, one such instance occurred in the first few hours when the traffic exceeded the upper control limit.

5.3 Visualizing Geospatial Information

Geospatial visualization can help us answer the following two questions:

* Where are events occurring?
* Where are new events likely to occur?

Location information of a Tweet can be identified using two techniques as explained in Chap. 2:

* Accurately through the geotagging feature available on Twitter.
* Approximately using the location in the user's profile.

The location information is typically used to gain insight into the prominent locations discussing an event. Maps are an obvious choice to visualize location information. In this section, we will discuss how maps can be used to effectively summarize location information and aid in the analysis of Tweets. A first attempt at creating a map identifying Tweet locations would be to simply highlight the individual Tweet locations. Each Tweet is identified by a dot on the map, and such dots are referred to as markers. Typically, the shape, color, and style of a marker can be customized to match the application requirements. Maps are rendered as a collection of images, called tiles. An example of the "dots on map" approach is presented in Fig. 5.9. The map uses OpenStreetMaps tiles and presents two differently colored dots. The blue dots are plotted using the location field in the user's Twitter profile, while the green dots represent geotagged Tweets.

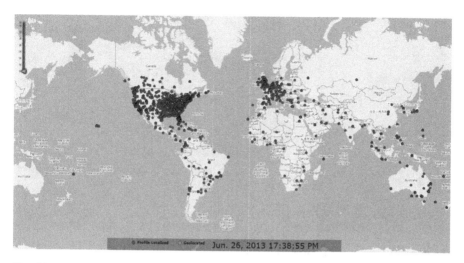

Fig. 5.9 An example of "dots on map" approach

5.3.1 Geospatial Heatmaps

The "dots on map" approach is not scalable and can be unreadable when there are too many markers. Additionally, when multiple Tweets originate from a very small region, the map in Fig. 5.9 can mislead readers into thinking that there are fewer than actual markers due to marker overlap. One approach to overcome this problem is to use heatmaps. In a geospatial visualization, we want to quickly identify regions of interest or regions of high density of Twitter users. This information for example could be used for targeted advertising as well as customer base estimation. Kernel Density Estimation is one approach to estimating the density of Tweets and creating such heatmaps, which highlight regions of high density.

Kernel Density Estimation (KDE): Kernel Density Estimation is a non-parametric approach to estimating the probability density function of the distribution from the observations, which in this case are Tweet locations.

KDE attempts to place a kernel on each point and then sums them up to discover the overall distribution. Appropriate kernel functions can be chosen based on the task and the expected distribution of the points. A smoothing parameter called bandwidth is used to decide if the learned kernel will be smooth or bumpy.

Using KDE, we can generate a heatmap from the Tweet locations, which is presented in Fig. 5.10. This figure clearly highlights the regions of high density and effectively summarizes the important regions in our dataset when compared to Fig. 5.9. Listing 5.9 summarizes the function *kernelDensityEstimate*, which can be used to generate a KDE estimate from the Tweets.

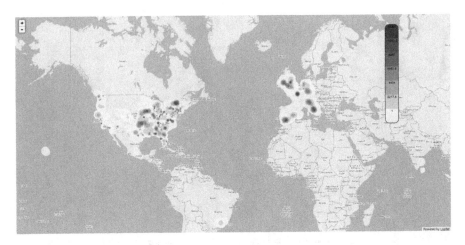

Fig. 5.10 An example of KDE heatmap on the map

Listing 5.9 Computing the KDE of the Tweet locations

```
kernelDensityEstimate: function(screenWidth, screenHeight, data,
    bandwidth, kernelFunction, distanceFunction) {
   // Step 1: Default to Epanechnikov kernel and Euclidean
       distance
   . . .
  //matrices that hold the points at various stages in the
      computation. Each will be the size of the screen (in pixels
      ).
    var pointMatrix = kernel_density_object.makeZeros(
        screenHeight, screenWidth, 0),
        bandwidthMatrix = kernel_density_object.makeZeros(
            screenHeight, screenWidth, 0),
        kernelDensityMatrix = kernel_density_object.makeZeros(
            screenHeight, screenWidth, 0),
        maxPoint = 0;
 // Step 2: Compute bandwidth matrix which stores the radius
    required to find <bandwidth> points at each cell
    for(var row = 0; row < screenHeight; row++){
      for(var col = 0; col < screenWidth; col++){
  . . .
  //Step 3: kernel matrix is the result of bandwidthMatrix pushed
      through the kernel function
    for(var row = 0; row < screenHeight; row++){
      for(var col = 0; col < screenWidth; col++){
  . . .
    //kernelDensityMatrix now holds a matrix of intensity values
        for each point
    return {
      'estimate': kernelDensityMatrix,
      'maxVal': maxPoint };
  . . .
}
```
Source: TwitterDataAnalytics/js/kernelDE.js

5.4 Visualizing Textual Information

Text is an integral part of Twitter. Here, we describe two approaches to visualize text.

5.4.1 Word Clouds

Word clouds highlight important words in the text. Typically, the frequency of a word is used as a measure of its importance. Word clouds are an effective summarizing technique. In word clouds, importance of a word is highlighted using its font size. The language used on Twitter is multilingual and mostly informal. Punctuations and correctness of grammar are often sacrificed to gain additional characters. Abbreviations are also frequently employed. To generate a word cloud, first we remove these elements and break the text into tokens. Then the frequency of each token is counted in the text using the method *GetTopKeywords*, which is summarized in Listing 5.10.

Listing 5.10 Extracting word frequencies from Tweets

```java
public JSONArray GetTopKeywords(String inFilename, int K,
    boolean ignoreHashtags, boolean ignoreUsernames, TextUtils
    tu) {
        //Read each JSONObject in the file and process the Tweet
        . . .
        /** Step 1: Tokenize Tweets into individual words. and
            count their frequency in the corpus
            * Remove stop words and special characters. Ignore
                user names and hashtags if the user chooses to.
            */
        HashMap<String,Integer> tokens = tu.TokenizeText(text,
            ignoreHashtags,ignoreUsernames);
        Set<String> keys = tokens.keySet();
        for(String key:keys) {
                if(words.containsKey(key)) {
                        words.put(key, words.get(key)+tokens.get
                            (key));
                }
                else {
                        words.put(key, tokens.get(key));
                }
        }
        . . .
        // Step 2: Sort the words in descending order of
            frequency
        Set<String> keys = words.keySet();
        ArrayList<Tags> tags = new ArrayList<Tags>();
        for(String key:keys) {
                Tags tag = new Tags();
                tag.setKey(key);
```

```
                tag.setValue(words.get(key));
                tags.add(tag);
        }
        Collections.sort(tags, Collections.reverseOrder());
        // Step 3: Return the first K words
            . . .
        return cloudwords;
}
```
Source: Chapter5/text/ExtractTopKeywords.java

To prevent information overload, we generally choose the top K words to create a word cloud using the method *DrawWordCloud* summarized in Listing 5.11. An example of a word cloud containing the top 60 most frequent words from the sample dataset is presented in Fig. 5.11. The word cloud effectively highlights the key events of the day, which consisted of mass arrests of protesters of the Occupy Wall Street movement by the NYPD in Zuccotti Park.

5.4.2 Adding Context to Word Clouds

Word clouds are effective in summarizing text. However, they place the responsibility of understanding the context of usage of these words on the reader. This is often not straightforward due to the limited information present in the word clouds. For example, if two words are used with relatively similar frequency, they are both highlighted equally in the visualization. However, a reader cannot determine if the words were used together or separately. This problem can be alleviated by using another dimension to add context to word clouds. Here we show how to use the time of usage of words to create a visualization with more context. To demonstrate this idea, we pick the top keywords observed in the word cloud in Fig. 5.11 and organize them into five broad topics as follows:

Fig. 5.11 An example of a word cloud containing the top 60 words

Listing 5.11   Creating word clouds

```
DrawCloud:function(words) {
        var fill = d3.scale.category10();
        d3.select("#cloudpane").append("svg")
        .append("g")
        .attr("transform", "translate(400,400)")
        .selectAll("text")
        .data(words)
        .enter().append("text")
        .style("font-size", function(d) { return d.size + "px";
            })
        .style("font-family", "Impact")
        .attr("text-anchor", "middle")
        .attr("transform", function(d) {
          return "translate(" + [d.x, d.y] + ")rotate(" + d.
              rotate + ")";
        })
        .text(function(d) { return d.text; });
}
```
Source: TwitterDataAnalytics/js/wordCloud.js

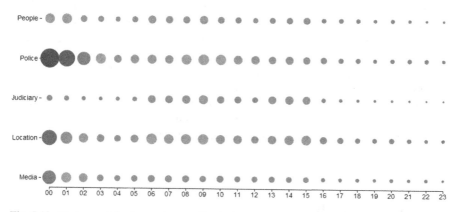

Fig. 5.12 Heatmap of the five topics combined with temporal information

1. **People:** protesters, people
2. **Police:** nypd, police, cops, raid
3. **Judiciary:** court, eviction, order, judge
4. **Location:** nyc, zuccotti, park
5. **Media:** press, news, media

The time and volume of usage of these topics is presented in a topic chart in Fig. 5.12.

Listing 5.12 Creating the topic chart

```
CreateTopicChart:function(json) {
    var r = Raphael("vizpanel");
    r.dotchart(10, 10, 1000, 500, json.xcoordinates, json.
        ycoordinates, json.data, {symbol: "o", max: 20, heat:
        true, axis: "0 0 1 1", axisxstep: json.axisxstep,
        axisystep: json.axisystep, axisxlabels: json.
        axisxlabels, axisxtype: "   ", axisytype: "|",
        axisylabels: json.axisylabels})
    .hover(function() {
                        this.marker = this.marker || r.tag(
                            this.x, this.y, this.value, 0,
                                this.r + 2).insertBefore(this);
                        this.marker.show();
                    },
                    function () {
                                    this.marker && this.marker.
                                        hide();
                    });
}
Source: TwitterDataAnalytics/js/topicChart.js
```

The topic chart can be created using the method *CreateTopicChart*, which is
presented in Listing 5.12. In the chart, the granularity of the time is set to 1 h.
Information is presented in the form of a heatmap, where both the color and the
size of the node represent the frequency of the occurrence of the topic. Police
related words are discussed more often than other groups. We also observe that
the discussion related to Judiciary does not gain traction until the middle of the day.
Time here can also be replaced by other dimensions such as the location, depending
on the intended application of the visualization. This visualization goes beyond just
the frequency of the words by leveraging the time dimension to help elicit interesting
patterns in the text.

5.5 Further Reading

Various visualization approaches presented in this chapter are organized according
to seven important data types discussed as part of the "task by data type" taxonomy
by Shneiderman [4]. The visualization mantra, which is used as the guideline for
building the visualizations in this chapter is also discussed in [4]. The principles
of graph drawing were proposed by Fruchterman and Rheingold in their paper on
the use of force-directed layout in graph drawing [3]. D3 uses the Quad-tree based
optimization technique proposed by Barnes-Hut to reduce the complexity of computing forces between nodes from $O(n^3)$ to O(n log n). Barnes-Hut optimization
aggregates the nodes into groups by storing them in a data structure called the quad-
tree. Each node of the tree then represents a region in the space and forces only

need to be computed between a node and a region if it is sufficiently farther away in the tree. Zooming and focus+context are popular techniques to make visualizations more useful. A review of the different uses of the techniques can be found in [1]. Additional information in Kernel Density Estimation can be found in [6]. The choice of a color scheme is crucial for the interpretation of the task. To represent the importance of the nodes in the network and density of the Tweets in heatmaps, we use a 7-class sequential color scheme from ColorBrewer 2©.[2] Guidelines for choosing the right color scheme can be found in [5]. The layout of the word clouds to visualize text is based on the popular Wordle system by Jonathan Feinberg [2].

References

1. Andy Cockburn, Amy Karlson, and Benjamin B Bederson. A Review of Overview+Detail, Zooming, and Focus+Context Interfaces. *ACM Computing Surveys (CSUR)*, 41(1):2, 2008.
2. Jonathan Feinberg. Wordle. *Beautiful Visualization*, pages 37–58, 2010.
3. Thomas MJ Fruchterman and Edward M Reingold. Graph Drawing by Force-Directed Placement. *Software: Practice and experience*, 21(11):1129–1164, 1991.
4. Ben Shneiderman. The Eyes Have It: A Task By Data Type Taxonomy for Information Visualizations. In *Visual Languages, 1996. Proceedings., IEEE Symposium on*, pages 336–343. IEEE, 1996.
5. Julie Steele and Noah Iliinsky. *Beautiful Visualization*. O'Reilly Media, 2010.
6. Tan Pang-Ning, Michael Steinbach, and Vipin Kumar. *Introduction to Data Mining*. Pearson Education, 2007.

---

[2]http://colorbrewer2.org/

Appendix A
Additional Information

In this appendix, we provide more information on building practical applications using the techniques discussed in the chapters of this book. In Sect. A.1, we discuss two systems built utilizing many of the techniques discussed in this book. In Sect. A.2, we discuss various academic and commercial systems built using Twitter data. In Sect. A.3, we provide more information on the libraries used to construct the examples described in this book and provide links to other resources.

A.1 A System's Perspective

Throughout this book, we have discussed techniques to build the necessary components for a system which collects information from Twitter and facilitates and analysis and visualization of the data. For those interested in examples of systems, which can be built using the techniques discussed in this book, we will present two systems, which demonstrate how the techniques can be used to build a solution to a real world problem.

TweetTracker© [1], is a platform to collect and analyze Tweets in near real-time. The objective of the system is to facilitate near real-time Tweet aggregation and to support search and analysis of the collected Tweets. In TweetTracker Tweets are organized into events to facilitate the study of related Tweets. Each event can be described as a collection of hashtags/keywords, geographic boundary boxes, and Twitter user IDs. Users create and edit events, which TweetTracker then uses to collect Tweets using the Streaming API. Tweets are indexed and stored in MongoDB using the techniques discussed in Chap. 3.

Analysis of the data is supported by various visual analytics. Temporal information of keywords and hashtags can be compared using the technique described in Chap. 5. Geospatial visualization is performed using the "dots on map" approach discussed in Chap. 5. Geographic location of the Tweets is obtained by using the contents of the Twitter profile location field in the absence of geotagging. Tweet text is summarized using a word cloud and a summary of the frequently mentioned

S. Kumar et al., *Twitter Data Analytics*, SpringerBriefs in Computer Science,
DOI 10.1007/978-1-4614-9372-3, © The Author(s) 2014

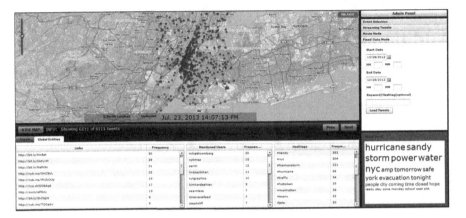

Fig. A.1 A screenshot of the main screen of TweetTracker. The figure shows the Tweets generated from the New York region during Hurricane Sandy. Tables below the map summarize the frequently mentioned users and resources in the Tweets

resources and people in the Tweets is also presented to the user. Boolean search capability is provided using a specific index built on tokenized text in MongoDB. Search is made flexible by the combination of text with other parameters such as the language of the Tweet and specific geographic regions. Figure A.1 shows an illustrative screenshot of TweetTracker.

TweetXplorer© [5] is a visual analytics platform to analyze events using Twitter as the data source. A user can analyze the data along four dimensions: temporal, geospatial, network, and text discussed in Chap. 5. A screenshot of the components of the system can be observed in Fig. A.2. User analysis of the Tweets is organized along specific themes comprised of keywords and hashtags. The network component is implemented in a fashion similar to the one described in Chap. 5. A user can zoom-in to a specific region to visualize the text in the form of a word cloud. The time series presents a global comparison of the traffic for the defined themes. The network component is also implemented in a similar fashion to the one described in this book. Each node is associated with a specific theme and internal node color instead of size is used to indicate the number of retweets received (darker colored nodes are retweeted more often). The network and map visualizations are also connected to each other. A selection in one causes the other to be automatically filtered to show the corresponding information.

A.2 More Examples of Visualization Systems

Building Twitter-based systems to solve real world problems is an active area of research. *Twitris* [6] is an online platform for event analysis using Twitter. The system combines geospatial visualization, user network visualization, and sentiment

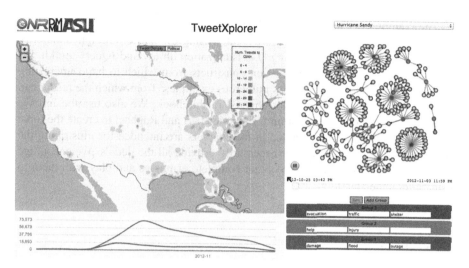

Fig. A.2 A view of the different components of TweetXplorer. The figure shows information pertaining to three themes

analysis to aid its users in analyzing events via different perspectives in near real-time. *TwitterMonitor* [4] is a system to detect emerging topics or trends in a Twitter stream. The system identifies bursty keywords as an indicator of emerging trends, and periodically groups them together to form emerging topics. Detected trends can be visually analyzed through the system. *TEDAS* [2] is an event detection and analysis system focused on crime and disaster events. TEDAS crawls event related Tweets using a rule-based approach. Detected events are analyzed to extract temporal and spatial information. The system also uses the location information of the author's network to predict the location of a Tweet when the Tweet is not geotagged. *SensePlace2* [3] supports collection and analysis of Tweets for keyword searches on-demand. The system focuses on three primary views: text, map, and timeline, to enable exploration of data and to acquire situational awareness.

A.3   External Libraries Used in This Book

All the examples in this chapter are written primarily using Java and open source libraries which can be downloaded at no cost to the reader. All the code samples discussed in this book can be obtained at the book's companion website *http://tweettracker.fulton.asu.edu/tda*.

The examples in Chap. 4 use a public network analysis library, JUNG.[1] Visualization examples discussed in Chap. 5 are created using JSP and JavaScript. A majority of the visualizations are built using D3 visualization library and JQuery toolkit. D3 provides a wide array of visualization constructs from which to build interesting visualizations. The library also has numerous examples[2] from which the reader can learn to build visual analytics not covered in this book. We also use the InfoVis toolkit in the network visualization for the pie chart and Raphael to create the topic charts. The code snippets included in the chapters are intended for illustrating the concepts and techniques and not necessarily provide all the details. We encourage the reader to visit the website to obtain complete examples and play with them to gain better understanding.

References

1. S. Kumar, G. Barbier, M. A. Abbasi, and H. Liu. Tweettracker: An analysis tool for humanitarian and disaster relief. In *Proceedings of the 2011 International Conference on Weblogs and Social Media*, 2011.
2. R. Li, K. H. Lei, R. Khadiwala, and K.-C. Chang. TEDAS: a twitter-based event detection and analysis system. In *Proceedings of the 2012 IEEE International Conference on Data Engineering (ICDE)*, pages 1273–1276. IEEE, 2012.
3. A. MacEachren, A. Jaiswal, A. Robinson, S. Pezanowski, A. Savelyev, P. Mitra, X. Zhang, and J. Blanford. SensePlace2: GeoTwitter analytics support for situational awareness. In *Proceedings of 2011 IEEE Conference on Visual Analytics Science and Technology (VAST)*, pages 181–190, Oct. 2011.
4. M. Mathioudakis and N. Koudas. Twittermonitor: Trend Detection Over the Twitter Stream. In *Proceedings of the 2010 ACM SIGMOD International Conference on Management of data*, pages 1155–1158. ACM, 2010.
5. F. Morstatter, S. Kumar, H. Liu, and R. Maciejewski. Understanding Twitter Data with Tweet-Xplorer. In *Proceedings of the 2013 ACM SIGKDD International Conference on Knowledge Discovery and Data Mining*, ACM, 2013.
6. H. Purohit and A. Sheth. Twitris v3: From citizen sensing to analysis, coordination and action. In *Proceedings of the 2013 International Conference on Weblogs and Social Media*, 2013.

---

[1]http://jung.sourceforge.net/

[2]https://github.com/mbostock/d3/wiki/Gallery

Index